About the Author

Dr. Kalluri Lakshmi Narasimha Murthy was born on 2nd September, 1949 in Vegeswarapuram. He has taken his M.A. (Geography) special in Agricultural Geography in 1973 from Osmania University, Hyderabad and had his Ph.D. (Regional Planning) in 1985 from Andhra University, Waltair, Andhra Pradesh.

He worked as Lecturer, Senior Lecturer and Reader in Geography from 1976-2007 mid April and also Principal (FAC) Government Degree College for Women, Srikakulam and Principal, Government U.G. and P.G. College for Men, Srikakulam, April 2007 to 30th September, 2007.

Dr. Murthy has presented research papers in the National Association of Geographers, Indian Annual Conferences from 1981-2012 and also presented a paper in the 27th International Geographical Congress (IGU, 1992), Washington D.C., U.S.A. Authored 3 books especially on "Survey of Research in Geography (1992), Geographical Research (1999) and Indian Social Science Research (2012) besides the present volume on "Research Methodology in Geography" (2013) also published research papers on Library and Information Science-Scientometric and Biblometric studies.

RESEARCH METHODOLOGY IN GEOGRAPHY

(A Text Book)

Dr. K.L. Narasimha Murthy

CONCEPT PUBLISHING COMPANY PVT. LTD.
NEW DELHI-110059

ISBN-13: 978-93-5125-021-0 (HB)

First Published 2014

Published and Printed by

Concept Publishing Company Pvt. Ltd.
A/15-16, Commercial Block, Mohan Garden
New Delhi-110059 (India)
T: 25351460, 25351794, *F*: 091-11-25357109
E: publishing@conceptpub.com, W: www.conceptpub.com
Editorial Office: H-13, Bali Nagar, New Delhi-110 015, India.

Cataloging in Publication Data--Courtesy: D.K. Agencies (P) Ltd. <docinfo@dkagencies.com>

Narasimha Murthy, K. L. (Kalluri Lakshmi), 1949-
Research methodology in geography : a text book / K.L. Narasimha Murthy.
p. cm.
Includes bibliographical references (p.) and index.
ISBN 9789351250210 (HB)
ISBN 9789351250227 (PB)

1. Geography—Research—India. 2. Geography—Research—Methodology. I. Title.

DDC 910.072054 23

Dedicated to **my mother, father** and to my **teachers** from Primary to University Education

[Late] Prof. B.N. Chaturvedi
[Late] Prof. V. Vidyanath
[Late] Prof. Mohd. Inamullah
[Late] Dr. K. Surender Reddy
[Late] Prof. Afzal Mohammad
[Late] Sri A.V. Krishnam Raju
[Late] Sri V.U. Bhaskara Rao
[Late] Sri M. Siva Nagaiah
[Late] Sri J. Ranga Rao
[Late] Sri A.V. Ramanujam
[Late] Dr. Y. Vittal Rao
[Late] Prof. G. Hanumantha Rao
[Late] Sri S. Anjaneyulu
[Late] Sri V.V. Narasimha Murty
[Late] Sri S. Jagannadham
[Late] Sri Ch. R.B. Sastry
[Late] Sri Ch.Nageswara Rao
[Late] Sri V.Veeracharyulu

Prof. S. Manzoor Alam
Prof. M.J. Subramaniam
Prof. K. Vithal Reddy
Prof. P.V. Sarma
Sri V. Narasimha Rao
Sri V. Mallapa Raju
Sri N.S.J.R.Varma
Sri K. Raghunadha Raju
Sri V. Upendra Rao
Sri N. Subbarayudu
Sri S.P. Butcheswara Rao
Sri K. Sivarama Raju
Sri P. Subba Rao
Sri R. Ranganadha Rao
Sri Y.V.S.N. Murty
Sri M.V.S.S. Lakshmana Rao
Sri A. Kama Sastry
Sri I.V. Sayanna
Smt. K. Prasadamma
Smt. G. Samuel

CENTRE FOR ECONOMIC AND SOCIAL STUDIES
Nizamiah Observatory Campus, Begumpet, Hyderabad - 500 016
Phones: 0140-23416610-13, 23402789, 23416780, Fax: 040-23406808
E-mail: post@cess.ac.in, Website: www.cess.ac.in

Foreword

Research Methodology in Geography by Dr.K.L.Narasimha Murthy is an important document involving intensive efforts to examine the current scenario of geographic research in the Indian sub-continent. The work has been divided into nine chapters. The First Chapter deals with an Introduction; Methodology in Geography is presented in the Second Chapter; and in the Third Chapter, Nature of Geographic Research has been described.

Design of Geographic Research – its objectivity and Geographic Survey is presented in the Fourth Chapter. Sampling Method and Experimental Method, Data Collection and Inputs in geographic research have been described in the Fourth, Fifth and Sixth Chapters.

The Seventh Chapter and the content of geographic research has been described such as (i) Research Methodology in Physical Geography, (ii) Research Methodology in Economic Geography, (iii) Research Methodology in Population Geography, (iv) Research Methodology in Urban and Settlement Geography, and (v) Research Methodology in Regional Geography, Geography of Resources, Regional Planning and Development. In Chapter Eight, Statistical Geography has been presented with a research paper and finally, Project Report writing and presentation of Research Paper in a National Conference has been described. The Ninth Chapter deal with a research proposal which is to be submitted for financial assistance has been presented.

This book provides wealth of reference material for the post-graduate students and researcher's in social sciences.

(Prof. S. Galab)
Director
Centre for Economic and Social Studies
Nizamiah Observatory Campus, Begumpet
HYDERABAD-500 016. A.P., India
Phones: 91-40-23402789, 23416780

Foreword

I am happy to write foreword to *Research Methodology in Geography* book, brought out by Dr. K.L. Narasimha Murthy, former Principal of Government Degree and P.G. College for men, Srikakulam Andhra Pradesh.

The significance and utility of Research Methodology in geography, have gained momentum by the introduction of Quantitative techniques in the twentieth century. Changes have taken place in the approaches due to the increased latest technology in place of the systematic and scientific studies in Geography. Geographical research on a large scale is taken place in many universities and institutions of India, following generalized and specialized set-Research Methodology.

In the second chapter of this book Dr. K.L.N. Murthy has presented the conceptualization of Research Methodology related to physical geography in association with space and time.

Methodology has been drawn to work with economic and regional resources for regional development which includes spatial development, agricultural, industrial and other responsible factors at National/State/Regional level. Evolved methodology for human, urban, rural, medical and health geographical studies.

In other chapters of the book, Dr. Murthy enlightened the different themes of nature of geographical research design, other techniques to be followed such as sampling and experimental methods.

The seventh and core chapter is divided into five subunits and emphasizes the detailed application of methodology to all the components explained earlier in chapter two. Eighth chapter recommends the methodology to follow statistical and cartographic techniques, statistical mapping and presentation of analysis mainly focusing the future scenario of the subject.

Ninth chapter deals with the different ways and styles of report writing/thesis writing. It also gives information regarding project preparation, research paper preparation etc.

I congratulate the author to bring out various facets of research methodology for various purposes mostly touching basic research in geography.

I hope and wish that this book will be found useful to all those interested in geographical research to produce a proper plan for concrete spatial development.

Prof. N. Nagabhushanam
Head, Department of Geography
S.V.U., Tirupati-2

Preface

"Research Methodology in Geography" examines the current research methods both qualitatively and quantitatively and its application to geographic problems. The entitled work further seeks to identify problems and its relevance to environmental and applied sciences is divided into the following sub-fields of geographical science and its vertical and horizontal growth over a period of time has been assessed. They are:

(1) Physical Geography, (2) Economic Geography (Agricultural and Industrial Geography, (3) Urban Geography, (4) Rural Geography, (5) Population Geography, (6) Geography of Health and Diseases (Medical Geography), (7) Regional Geography, (8) Regional Development and (9) Current Status of Geographical Science.

'Statistical Methods' in Geographical Science is an attempt to project and present important geographical problems and also to find solution to complex geographical problems, besides the evaluation by statistical methods and mapping of geographic variables through different cartographic methods is an important aspect of this venture. The prime aim or attempt of the author is to identify, recognize and present the role of technology such as remote sensing Geographic Information System (GIS) and GPS (Geographical Positioning System besides aerial photo information.

According to V.C. Jha "The importance of geography as an academic science as well as its role in National Planning has been widely recognized in the developed world. In India, it has yet to receive the recognition. The practitioners of geography in India have made significant contribution in regional planning, urban planning, watershed management, resource management environmental management, census operation, cartography, land-use mapping, anthropology, rural development, terrain evaluation for civil and military purposes, highway and settlement planning, route location and site selection, etc."

"Geography now plays a multidisciplinary and interdisciplinary roles in nature, with the result that planners, engineers, exploration geologists and geophysicists, soil scientists, hydrologists, demographers, cartographers and environmentalists have also developed an aptitude for geography, mainly because of the practical application of geographical knowledge in their respective disciplines" [V.C. Jha].

Increased food production in tropical countries is essential to feed rapidly growing population (Asian experience) identification of agriculture has resulted in drastic changes in land-use patterns, directly affecting the denudation processes. Land degradation and desertification have become common problems due to climatic conditions and changes and anthropoid impacts through land-use particularly in the tropical lands. It is understood that land degradation processes that affects the world's landscapes. Sustained food production can only be maintained, if these problems are identified, understood and solved.

Spatial information systems have opened up a new vista for the geographers by providing multifaceted information, at a short notice about the earth's surface which happens to be the home of men. India has also come a long way in meaningful application of remote sensing technology by developing its own remote sensing satellite system which provides spatial, spectral and temporal information about the globe.

It is essential to evaluate the research work and innovative informatics related to the fields of Geography, climatology, Hydrology, Oceanography, Biogeography, forest ecology and environmental preservation, economic geography, land-water resources, land degradation and desertification's, population and settlement geography, agricultural and industrial geography, urban geography, multilevel planning, geography of mountains, social and cultural geography, political and medical geography, Anthropology and Tribal geography or geography of tribes, Historical geography, Rural Development and Regional Planning, Remote Sensing, GIS and Thematic geography and spatial information technology, are the focal themes of Indian geography based on remote sensing and spatial information technology.

Dr. K.L. Narasimha Murthy

305, Manikanta Sai Sudha Apartments
Meena Café Lane
Warsiguda
Secunderabad – 500 061
Andhra Pradesh (India)

1st September, 2012

Acknowledgements

The author expresses his deep sense of gratitude to Professor B. Chakravarthi, National Institute of Rural Development (NIRD), Rajendranagar, Hyderabad and also to Professor. V. Madhava Rao, Head, C-GARD, NIRD, Rajendranagar, Hyderabad for the encouragement intensively extended to the author during the year 2011.

The author is indebted to his teachers, Professor S. Manzoor Alam, Osmania University, Hyderabad, Professor. R. Vaidyanadhan, Andhra University, Visakhapatnam, Professor. P.V. Sarma, Department of Economics, Andhra University, Visakhapatnam, Professor (Late) N. Balakrishna Reddy, S.V. University, Tirupati and Professor (Late) E. Swaminathan, S.K. University, Anantapur.

Dr. Murthy was invited by Professor M. Sambasiva Rao, Department of Geography, S.K. University, Anantapur, three times to present thematic papers on "Environmental Degradation and Development" during March 2009, March 2011 and September 2011. Professor R. Ram Mohan Rao (Retd), Department of Geography, Osmania University, Hyderabad, who has been giving guidelines from time to time to the author.

The author was permitted by Professor S. Galab, Director, Centre for Economic and Social Studies (CESS), Hyderabad for critical appraisal and also to consult Library and for the help extended by Mr. J. Simhachalam, Asstt. Librarian, CESS, Hyderabad and to Sri A. Srinivas for typing the entire manuscript and preparing the C.D. on time. The author is highly indebted to them.

Contents

Foreword by Prof. S. Galab vi
Foreword by Prof. N. Nagabhushanam vii
Preface ix
Acknowledgements xi

1. Introduction 1
2. Methodology in Geography 3
3. Nature of Geographic Research 17
4. Design of Geographic Research – Objectivity – Geographic Survey 48
5. Sampling Method and Experimental Method 53
6. Data Collection and Inputs in Geographic Research 68
7. Research Inputs (Data Collection, Processing and Results) in Major Branches of Geography – The Ongoing Process : An Overview 82
8. Statistical Geography 169

8.1. Research Paper: Population of India: A Geographical Appraisal and a Statistical Account 187

9. Guidelines for Project Report Writing 204

9.1. Research Proposal : Nutritional Deficiency Diseases and their Prevalence: A Case Study of Urban Pondicherry 212

Bibliography 216

Index 219

1

Introduction

"Geography" is defined as the spatial distributional study of resources and development. The sources of energy are land, water, sun, air and *Akasahaa* means the study relating to *Panchabhutas* land form (building and degradation) water resources – (*Apahaa*) = (cold currents and warm currents), *Teja* indicates and relates to climate (low and high temperature, Rainfall (high and low) precipitation, etc., *Vayu* relates to air (warm and cold – study of air masses and *Akasahaa* (the blanket of air encompasses from land to sky, i.e., the sum total of the resource study upon the surface of the earth and water resources on the ground and underground (Hydrologic pattern) and also the existing climatic phenomena upon the landforms and also upon the oceans are important aspects of geographic science. This typical study relates to Geography and its relevance to environment and allied sciences have its own impact and imprint typology to understand the birth/source of phenomena and also its decay. Over the land and water under the surface, with terrific attitude in building up of terrain, its functionality and variability in different environments is an important factor of study and research by geographers and geoscientists.

Physical geography or physical basis of Geography/principles of physical geography relates to the study of physical principles and patterns of physical structure and structural levels leaves an indelible mark upon the surface of the landform. Physical geography is mainly sub-divided into three sub-fields which have its individual claim and relevance as Physical Geography. The major sub-fields of Physical Geography are (1) Geomorphology, (2) Climatology, (3) Oceanography and the matter/substance ranged into three major forms, i.e., *Pruthivi* (land) and its associated form (High, Medium and Low according to height from the Mean Sea Level (MSL). Geomorphology is the study relating to landforms and this is a systematic and scientific study of landforms and associated factors relating to origin, growth, development and decay. In simple terms of terminology as stated by Geomorphologists and has been identified and recognized as a cycle such as Fluvial cycle (River cycle) exogenic, underground water (Endogenic, i.e., pattern); Aeolian cycle (Air as an agent of erosion, transportation and deposition); glacial cycle and marine cycle, etc. It is apt to examine and say that land forms are subjected to typical erosion, transportation and deposition and it is further to examine that exogenic and endogenic factor's influence have combinedly results to deformed structure to the original landforms and attains new typical structure (it is due to erosion, transportation and deposition) and finally deposits the eroded material which has been placed at a new place far away from its original place. All this process depends upon the physical process, chemical process and biological processes. Mechanical and chemical weathering and biological processes helps to landform (aggradations and degradation) structure. The following study added to the knowledge as Research addition to the existing knowledge and the methodology in geography is presented in chapter 2.

2

Methodology in Geography

1. Methodology

The principal aim of the author is to present methodology in geography in this chapter. Geomorphological studies include the following themes and they are important as an additional inputs to the existing literature.

(1) Land Slides Hazard: Typology, mechanism and mapping are important inputs in research.
(2) Management of water related disasters — particularly to mountain regions.
(3) Study of litho structural control on morphometric parameters.
(4) Landscape ecological planning and management.
(5) Development of hill slope profiles on the rocks.
(6) Production of Ungauged Basin (PUB); Hydrological challenges.
(7) Land degradation/desertification as an environment Dilemma.
(8) Management of surface water resources.
(9) Loss of forest cover and land degradation.
(10) Remote sensing application in change detection of Ravine land.
(11) Desert National Park and Unique Biosphere Reserve for conservation and development of biodiversity.
(12) Forest cover change.
(13) Physical environment — study through remote sensing and Geographic Information System (GIS)

(14) Assessment of Drainage Basin morphometric patterns under varying terrain units.

Climatology occupies an important place in Physical Geography. From local to global, Climatic variables have variations in temperature, rainfall, aridity and the associated elements such as air pressure ranges from low, medium and high. Koppen's climatic classification is an important and noteworthy classification and it has its applicability throughout the globe. Through Waite's climate classification adds more to climatology. It is observed that over a period of time, climate changes rapidly and has its influence on agriculture and agricultural development. Agro-climatological studies are important and micro climates play an important role in agriculture and agricultural operations. Mountain/plateaus/plains (hierarchical pattern of landforms from the Mean Sea Level) [MSL] have its impact of climate change on water resources and sustainability in Mountain/ Plateau/plain regions with special reference to Indian climatology, Indian climate has its varied and vivid influential phenomena on landforms. Moisture sufficiency and deficiency are important as determining factors of crop patterns. Impact of Rainfall on Agricultural productivity/crop is a major study which could be related to crop planning, keeping in view of moisture sufficiency/deficiency. Pedogeomorphic surfaces on land-use pattern which has direct/ indirect influence of climate (local climates) and has a bearing on population distribution. This pattern could be visualized in a profile section.

The extremes of climate could be visualized in the form of drought and floods. It could be identified that high temperatures clubbed with low rainfall all leads to typical drought, whereas high rainfall clubbed with low temperature leads to a typical flood. Droughts and floods are considered as natural disasters and have its effect on agricultural crop productivity. Rainfall/temperatures/aridity/and soil moistures play an important role in regional agricultural land-use and crop behaviours. With reference to the above, it is identified that watershed development planning for sustainable agriculture including micro-watersheds in India are important and occupies pivotal role in agricultural crop planning.

Government of India/respective State governments should have a disaster management cell in the ministry, keeping in view of the frequent occurrence of floods and droughts. It helps to mitigate the loss due to natural hazards particularly in India/different States. It is further noted that the delineation and zonation of flood prone and drought prone areas in the State/group of States in India are much awaited. Floods are considered as a threat to sustainable development. Human settlements/civilization develops according to climate and climate change is also subjected to changes in settlement system. Global warming implications, causes and concerns are directly have its effect on human mobility/settlement system.

Air quality decreases to the increasing distance of metropolitan/ cosmopolitan cities. It is due to pollutants, the quality of air decreases. In the settlements abutting coast and the urban air quality and its influence on people residing in that particular zone and are eventually creates health hazard. In the same way, PH of rain water also varies from decade to decade.

Hydrological regime/tornadoes and water spouts/rainfall trends and underground water regimes are important in deciding the settlement system whether it is permanent, semi- permanent/temporary character in flood prone (drought prone/earthquake zone micro climate and hydrological changes are important study leads to agro-settlement system).

Environmental impact studies relating to geomorphic, climatic and oceanographic locations, site and situation invariably present changes in landform and land-use depending upon climatic condition and environmental characteristics and also waterbodies (oceans, seas, coasts, beaches) and the study leads to coastal geomorphology and the associated structure/process and stage elements over space is an important space-time study to regional geomorphology and development.

2. Economic Geography

Economic geography which studies the economic activities of man upon the earth and it is called as economic geography "If we observe the human societies in different areas on the surface of the earth and people's customs, food, means of life, natural and cultural aspects, we notice great variety. The economic activities, cultural and technical development of the people living in a particular area affect economic and human life to a great extent American scholar consider it a link, between economics and geography, which is called geo-economics or someone called it geonomy (Geography+Economics)". B.S. Negi,

Economic and Commercial Geography of the World, published by Kedarnath, Ramnath, Meerut defined ("Economic Geography" is the study of economic activity that is being activated by human being for earning livelihood for sustainable development that is to state that harnessing of economic resources (both agricultural and industrial) for human development through the skills of human being" (Quotation Mine please).

Spatial distribution of resources which vary from place to place, i.e., resources are not uniformly distributed upon the earth's surface, as a result, spatial disparities in resource endowments, further leads to human uproar, finally, leads to sparse development with variable densities, which promotes poverty and hunger. Harnessing of regional resources leads to regional development, which, people have its major share in developing the economy over a period of time (Five Year Plans in India) for achieving desired development.

"Economic Geography" is an important branch of Geography and it is the study of economy or economics by geographers. Economic resources promotes to agricultural development and industrial resources in turn promotes for the further cause of industrial development. Agricultural development and industrial development may depend upon the policies and procedures of the Government of India (GOI), New Delhi, as well as its individual States and U.T. (Union Territories).

If we examine any State or Centre, in depth, one could identify the regional disparity in work participation [occupational structure as well as skilled and unskilled workers in an economy]. In the same way, geographic mosaic of culture, environment and human development leaves its indelible mark on economy, Industrialization and also urban industrial growth. It is a typical deciding factor for development. In the agro-sector, it could be observed that 'dairying' also helps as an instrument of employment generation.

"Micro level" studies may help to understand any planner to identify and recognize socio-economic and agricultural situation at village level/cluster of villages/*mandal*. The impact of government policies and use of conventional energy helps to understand the developmental activities and also to identify the trade links, i.e., rural to urban and urban to rural. Resource development helps to urban fringe and urbanization, which depends upon the strategy of development of planning.

'Emigration' and 'immigration' are the two powerful agents of growth promoting/growth retarding factors for the interaction of people and commodity from one place to another place. That is to say that identity, labour market and conflict-political economy of labour immigration is a process for the upliftment of an economy.

'Biosphere' play an important role in the development of economy of any State/nation/country in the world. 'Biodiversity' helps to development. In this direction, the governance should be able to formulate trade policies for its growth. For example sustainable rural development, fodder and pasture development, systems of vertical horticultures, etc., and international seminar/conference on bio-diversity (conducted in India, October 2012) assumes importance in global development.

"Changes in land-use and cropping pattern" is possible if the technology is identified in tune with soil/pedalogical factors. Another important factor, such as power projects/river water project (multi-purpose projects) may help for the alleviation of agricultural system from dry agriculture to wet agriculture, this leads to changes in land-use and cropping pattern besides occupational structure of people. Development and displacement of ethnic population is due to installation or augmentation of multipurpose river projects/power projects and it leads to geopolitical stability or instability in a political unit/units.

'Food security' for the growing population is a dire need to adjust with the environment. Fast growing population should safely be guarded with water, food and shelter security. Keeping in view of the instable conditions in agricultural prosperity, natural disasters often play an important role at the national/State/district or sub-minor level units in a country. Regional patterns of food security is a powerful planning operation.

'Locational theories' play an important role in the regional economic system. 'Theories of location' are different and may not uniformly fits in and it varies from environment to environment. The application of location theories and its critical appraisal may vary from time to time. Space and time, may also differ. The Western theories may not fit in Indian site and situation totally/partially for development.

'Urbanization' is a powerful factor which contributes to the economy and generates employment in various sectors to people. It has an impact on environment. Spatio-temporal variations could be identified in different area levels for the urban-rural differences and it may cause further differences in human resource development.

Trends of agrarian structure, patterns of land holding structure and creation of infrastructural facilities leads to the development of agriculture and allied activities. The above factors are considered as growth promoting factors to agriculture. Identification of agricultural resources and planning for development is a strategy for any administrative unit/macro, meso and micro level units leads to strategic development.

Agricultural modernization as well as development of social sector will help for economic development and technical efficiency in crop productions may promote farm production and productivity. Economic reforms may increase social tension among the people. The existence of disparities leads to disparities in the levels of agricultural crop production and productivity per hectare may also vary.

'Micro level studies' promotes in the increase of crop productivity and also to understand the phenomena of micro-regional structure. Micro level planning may help to integrate development in the sectors of agriculture, industry and social sector. Economic infrastructure and socio-economic development may be feasible if we plan the resources through a time bound and budget plan of operation.

Optimum utilization of resources leads to optimum development, i.e., maximum benefits could be derived with a minimum cost factor which play an important role in the development of agriculture and agri-operations, land transformation in relation to distance is to be taken up for analysis.

Regional inequalities and regional imbalances lead to regional imbalances in food security. Regional crop yield variation besides land-use intensification and diversification for sustainable agriculture is a powerful strategy for regional development. The economics of crop productivity depends upon economic resource use and diversification of agriculture promotes to achieve higher profits.

Keeping in view of the above situation in mind, the planner should be able to identify the spatio-temporal patterns of foodgrain availability for agricultural development whether it is at regional/ State/national level. The prime aim of a planner is to identify regional disparities and plan spatial pattern of agricultural development and industrial development to iron out such disparities, a well thought plan should be augmented at the national and State level.

3. Urban Geography

"Urban Geography" contributes to regional/national development, leads to the utilization of resources (urban) and thereby creating urban infrastructure for urban socio-economic development. The facets of urban space and their variation should be studied in depth which further helps the planner to plan for urban development. "Urban Space" is an important place to create or inject urban functions at different area levels for uniform development. The levels of human resource development vary according to urban-rural differences in the total compositional structure of population. The urban planner should have to identify the causes for the variations in a regional set-up/regional structure in order to plan for development. Quality of life, gender bias, fertility behaviour in labour areas may differ from time to time over space even with in the urban zone. The urbanization process may change the city and city environs speedily. Urbanization depends upon the urban resources, environment and mode of operation of urban plan for development.

The classification of urban settlements leads to a pattern and process of hierarchy inter-district disparity leads to development in social sector, identification of social need, and social planning leads to socio-economic development of urban spaces. Sometimes urban green space may help for urban sustainability. Hierarchical market locations or market hierarchy may invariably help for regional development by utilizing resources to the optimum level.

The sub-regional studies such as inter State/inter district analysis may help for human development. Variations in human resources leads to a disparity in human development and the construction of human poverty index ultimately helps to identify the spatial variations and may be further utilized for designing the planning studies for development. The interaction of rural population with the market centres and their density in turn arises to develop a strategy for sustainable development. The role of market centres may give clue to understand the 'hierarchy' of market centres.

The study relating to the gender status and disparity at the regional level may be helpful to understand the women problems particularly those reside in urban slums. Human deprivation and development studies may be capable of identifying the core problems of the macro or micro level (A State level/A district level) or group of districts forming a block and spatial development strategy would be a suitable measure to iron out such deprivation. Class 3 and Class 4, towns sometimes present typical problems such as female work participation and their role in spatial development. Among social classes, it is pertinent and visible that variations in the level of poverty is an important aspect and the urgency of such inequalities should be mitigated. The role of urban planner waits for finding solution in this direction. Regional inequalities and disparities may affect the growth and development in a typical urban space. Sometimes they present urban hunger due to urban poverty.

Steps towards structural transformation at the national level is possible only through the growth of service sector. Impact studies such as impact of urbanization on land-use in such urban areas of a developing country. Urbanization process may help for the transformation of land-use from agriculture to non-agriculture, such land is being used for new settlements. This situation is visible in an urban fringe area. Urban government or governance may find solution in new spaces and interface dissimilarities.

Finally, the solution for the above problems is the implementation of integrated urban development strategy over a period of time may prove success and this rests only with the urban governance.

4. Rural Geography/Settlement Geography

"Rural areas" are the potential areas which generates or gears up development either for rural development or semi-urban fringe development. "Rural resources" should be transformed and transferred into urban activity and the sub-urban areas are added continuously to urban metropolitan areas or cultural transformation is on the online process and thereby agricultural activity is seen in a discontinuous way. We may conclude that welcoming to urban activities and permanently saying good bye to the rural activities should be a norm.

The current research in rural geography is ranged and presented thematically. The central research on rural areas are as follows:

(1) Study on urban and rural differences and identifying the sources and find meaningful solution to both of them.
(2) The urgent need of an urban geographer/urban economist/urban sociologist is the method of describing geographic association.

(3) To identify the place names and environment of rural settlements and sustainable rural development.

(4) Development of social sector — A study in inter district/inter-State disparity.

(5) The study relating to agricultural modernization among farmers gains momentum.

(6) Nature and extent of disguised employment study in drought/flood effected areas at nation/State/district level is to be taken up. A micro level study such as intra-Mandal and inter-Mandal may present disparities of development. An in-depth study may be helpful to design a strategic plan for development of rural areas.

5. Population Geography

'Geography of Population' or Population Geography' is a major branch of Geography and the study relating to geographic conditions and its environmental impact on spatial patterning of population and population structure is equally important. The structural composition of population belongs to a region/State/nation play a vital role and is considered as a resource of the region for strategic development. 'Environment' is a key to play an important role in the peopling of the region. The economic planning and development depends upon potential population (15-59 years) of the State/region, etc. The geographical conditions for the region/growth and development is specially and spatially examined in the following text of the geographic description sequentially 'migration' is a powerful tool, through this, people may shift from the original place of birth to child place to find employment/study/marriage. This social process creates instability in social situation in an innovative society where technology is offered for instantaneous development over a period of time. The organization of polity has its impact on migration process, whether it is immigration/emigration. The so called developed countries/nations may encourage social immigration so as to achieve development in a time bound and budgetary bound plan. Generally, people may migrate with skills from underdeveloped nations to a well developed and technological nations. For example, the migration from Asian countries to Americas/Africa/U.K. and European countries for employment/education or any other reason for migration.

Regional development reflection issues and policy in the third world attracts more to the demographers/sociologists/economists and geographers to undertake studies relating to migration. Due to construction of power/river projects, it is definite to migrate the population from construction to the safe areas and new livelihood patterns may be adopted. It is a dire need from the angle of migration. It is further observed that development and displacement of ethnic population from its original place to a new place which migrants may not show much interest to move and they cannot be detached from the existing attachments. They feel identity, labour market and conflict and political economy of labour immigration.

Regional disparity in work participation and work participation rates may differ from place to another place, keeping in view of their newness in their livelihood patterns. 'Occupational structure' sometimes changes the behavioural patterns of population. 'Dairying as an instrument of employment generation in a new settlement. Population movement (external to internal) may cause tension due to the disturbance of settlement and occupations. Regional geographic analysis of population pressure on land may also consider as one of the causes and it reflects land-man ratio. At the national level enterprise, growth and inequality in the rural economy may be observed and studies relating to growth depends upon the enterprise which they eventually be taken up.

Rural non-farm sector in India and its contribution to employment and income adds to the economy. Urban occupations may support a higher level of population is visible in any country and India is no exception to it and a case study in this direction. "Impact or decentralization policy on occupational patterns and mobility in tribal areas is much invited and awaited in order to identify the life patterns and livelihood system besides settlement system including women employment.

Even though, the agricultural amendments are made frequently to change the child labour, patterns of the child labour are visible at the national and State level. Spatio-temporal analysis of demographic condition and resource crisis is much invited to undertake such studies. Statistical models for finding out determinants of potentiality to get absorbed in jobs for different categories of labour in Indian States/ district (micro level) study may be much resourceful and human resource is utilized through their skills. A micro level analysis-status of women and their social interaction in an economy leads to women employment, employment and employability. Women should be given equal opportunity in all respects of development planning.

6. Geography of Health and Diseases — Medical Geography

'Health' and 'diseases' are nothing but they are the two faces of a coin and human races are prone to ill health which leads to disease. Environmental impact on people (human) and its effect is an important aspect of modern geographical studies. How the diseases will spread over and spatial variability in its dimension is the past/current trends of study in geography of health or health geography. Health care facilities, government (central/State) policy and creation of health infrastructure to the existing population in a district/State/nation which adds as an important aspect contribution to health geography. The post-health care facilities and present health care facilities have a telling effect on human population and settlement system. Keeping in view of the increasing population, there should be an increasing quantum in the infrastructure as a 'service provider' and it alleviates the quality of life of population. The increasing trend of infrastructure should be able to match the medical needs of the living population as well as the future prospects of population is based on the health care system. 'Urban medical care' has been increasing both vertically, and spatially. Corporate medical services are available to the urban population whereas rural population have been facing many inconveniences to reach the good medical care systems. 'Population' as norm and 'area' as norm should be taken as a deciding factor for providing medical care facilities. If the government (centre/State) is able to provide medical facilities to the growing population by allocating necessary funds in the Five Year Plans and the unutilized funds should be utilized systematically over a period of time to achieve development in health sector. 'Spatial medical care system' should be developed on 'integrated norm' and integrated planning may prove successful in medical care system.

7. Regional Geography/Regional Resource Evaluation/Regional Development/Rural Development

An Evaluation and evolution of sea change – Disparities in the levels of development is visible in the following paragraphs.

'Region' is defined as, "that part of the earth's surface which possesses the natural characteristics with administrative boundaries sometimes with variations which are bounded by natural physical boundaries intend/awaits planner's planning process with a well defined time bound objectives subject to earmarked funds for plan implementation for development" (quotation mine please). Geography of a region or regional geography is considered to be an integral and important synthesis to know the resources and waits for evaluation and development by a planner/geographer/economist/or a social scientist. Regional development is possible only through a systematic plan relating to rural and urban development. The evaluation/pilot study is quite helpful in generating the evolution and identifies the sea change in development studies.

"Disparities" in development lies in the resource availability for sustainable development to iron out such development disparities, it is imperative to implement a well conceived and defined plan with specific aims and objectives, which further helps for the eradication of disparities, though it is physical, nothing is lost, it is understood that social and economic disparities should not persist after the implementation of plan for social, economic and cultural development. Important regional studies were conducted by C.D. Deshpande (Pune); R.L. Singh (Varanasi), Ahmad, E. (Patna), V.L.S. Prakasa Rao (Delhi), L.S. Bhat (ISI, New Delhi), Moonis Raza (New Delhi), Manzoor Alam (Hyderabad), Vidyanath. V., and R. Ram Mohan Rao (Hyderabad), Sarma P.V. (Visakhapatnam), C.R. Pathak and R.N. Chatopadhyay (Kharagpur) during 1964-2000. Jasbir Singh (Haryana), M. Shafi (Aligarh), R. Singh (Bhagalpur), S.P. Chatopadhyay, National Atlas and thematic mapping organization, Kolkata, B.N. Sinha (Bhubaneswar), D.K. Singh (Bhubaneswar), A. Ramesh (Madras), P.S. Tiwari (Madras) and R.P. Misra (Mysore and Delhi) have contributed important studies intensively over an extensive geographic area for the cause of Indian Geography (specifically to various (major) branches of Geography over a period not less than 60 years which is from 1950 to to-date.

8. Marketing Geography or Geography of Marketing/Geography of Retail Trade, etc.

The study of rural settlement types and spatial typology play a central role in settlement geography cultural factors, social factors and economic factors will also contribute to a large extent in the spatial patterning of settlements and settlement system. Rural settlement types first studied by Enayat Ahmad (1952), "Rural Settlement Types in the Uttar Pradesh (United Provinces of Agra and Oudh) published in Annals of the Association of American Geographers, XLII, p. 233-46 (I, III). During 1967, B.J.L. Berry (1967) studied, "Geography of market centres and retail distribution", Engle wood Cliffs, New Jersey (III, IV), Berry, B.J.L. and Garrison, W.F. (1958), "Recent Development of central place theory" papers and proceedings of the Regional Science Association, 4, 107-20 (II) are important continuations to this branch of geography.

The spatial patterns and temporal synchronization of periodic market, *Swansea Geographer*, II, 15-28 (III) and it is understood that marketing is one of the vital missing links in the rural areas; yet few studies have been made of this vital aspect even during the present euphoria of rural development. Marketing geography should have to cover the aspects such as the theoretical base, spatio-temporal characteristics, hierarchy, functional and behavioural aspects into a pattern that reflects the micro as well as macro characteristics, draws comparisons with other conclusions and relates to the whole thing to the prime objects of the development of people in rural areas. It could be observed that vertical trade dimensions may restrict and generation of wealth is impaired leading to the perpetuation of backwardness.

Marketing geographic studies will stimulate further research and provide the planners that helps in a big way for formulating effective strategies to combat rural poverty. The important and pioneering study authored by Reed, W.J. (1931) entitled, "The Law of Retail Gravitation, Kvicker Bocker Press (II) and Railly, W.J. (1969), has prepared the methods of study of related relationship, Bureau of Business Research, The University of Texas, Austin (Reprint of Research Monograph, No. 4) (V). The Indian Contributions made by Shrivastava, V.K. (1976), "*Market Periodicity and Centrality*: *A Methodological Approach*, Mainzer Geographischer Studien, 10, 33-38 (II, IV) are important contributions to geographic literature.

The theoretical base for the entire studies were deduced from Christaller's Central Place Theory and other available theories of location. The location of a place has its prime importance in the regional settlement system or urban system or rural system of rural spatial settlement systems relating to the geography of marketing and periodic market system. The 'hierarchy' of central places play an important role in the total system. The hierarchic classes are divided into three tier system such as central villages, Service centres and growth centres. It is examined that the western theories which are governed by certain principles and environment forms a system relating to market/settlement system into "hexagonal pattern". The Indian environment play a vital role in the regional central place system. Indian studies proved that central places are formed into square type or triangular type, in some cases it forms into semi-hexagon in (Ref. Murty, K.L.N. '*Planning for Integrated Area Development*: *A Case Study of Tanuku Taluk, Andhra Pradesh*, Murthy, K.L.N., "*Ecology, Environment and Development* (2003) and Murty, K.L.N., "*Applied Geo-informatics*" (2007), wherein, the study proved that central places are formed in a ribbon type followed the transport lines in a backward area, whereas, in a delta, the central places are formed into triangular/semi hexagonal shape. Shape and size of settlements may play an important role in a settlement system depends upon the total environment.

It is concluded that major branches of geography have been offering thematic studies relating to search and research from the time immemorial. Major branches of geography in India have a systematic scientific role and place to conduct intensive research. An attempt is made by the author to identify and form formulations leading to hypothesis and its testing in Chapter 2; keeping in view of the existing body of knowledge. The role of technology in current geographical research such as GIS, R.S. and GPS are dealt separately in the seventh chapter. The nature of geographic research have to be examined and the same is presented in Chapter 3.

3

Nature of Geographic Research

'Geographic research' is nurtured in nature as a plant kingdom upon the earth that is to state that 'environment' is a powerful factor and decides the geographic process. "Geographic features" *versus* climatic impact on landforms and climatic impact on seas and oceans have a predominant role in the nature of geographic research "cause" and "effect" are considered as the two blades of a scissors, like demand and supply lines to intercept at a point which decides the price (in economics), climate's impact on geographic landforms particularly the impact of rainfall on agricultural production and productive levels in deltas, plateaus, hills and mountains. It is also observed that the influence of climatic factors such as rainfall (high and low), temperatures (high and low) and precipitation and aridity/arability upon land and waterbodies (seas and oceans) and air (cold and warm) extensively having a bearing on land, water and air, i.e., landforms, seas and oceans and blanket of air (lower and upper surface of the air from the mean sea level (MSL).

Highest landform among the major landforms from the mean sea level (MSL) is the mighty Himalayas; whereas the deepest of the ocean deep among the ocean deeps, is Mindnova *sea of Azores* and the temperature upon the rocks on the çelestial body—the moon—are the primary aspects of "scientific research" in physical geography/physical basis of geography which ejects out to geomorphology, oceanography and climatology in a non-scientific but crude form described as Ghana, *drava* and *vayu rupa,* i.e., lands, water and air.

From the time immemorial, the research activities in geography have been in the form of investigators travel, expedition, research, search and discovery. They had the unquenchable thrust for knowledge that is to state that they had inquisitiveness to learn new things, acquire new knowledge to grasp newness to the existing knowledge through their travel, expedition and innovative results to communicate to the society. The governance of different kingdoms had encouraged the travellers belongs to further expeditions and explorations.

Scientific geographic thought has resulted new expeditions and explorations from the earliest period to the present day. Among them are Aristotle, Strabo, Copernicus, Ptolemy, Ratzel, Ritter, Vidal de Blash, and their scientific explanations are still noteworthy and helps to the community at large. During that period, the investigations relating to the size of the earth, area of the earth, longitudinal distance from north to south pole, and the latitudinal distance (i.e.,) air performance on the zero latitude has longest distance and the distribution and distances from different angles were measured in miles and nautical miles (miles on the land and nautical miles upon the oceans) by major scientific community. Columbus discovered America. This is to say that Columbus has the courage, encouragement, efficiency and sufficient travel knowledge and as a result it so happened.

Keeping in view of the above, geographic research primarily belongs to observation and measurement. Observation by the physical eye and through instruments such as telescope, and measurements were based on mathematical principles and mathematical formulae helped to surveying and mapping of landforms. We conclude that search and research belongs to geographical science is based on mathematics, astronomy and physics. An attempt is made by the author to present the concepts such as (1) scientific method; (2) nature of Geographic/Social Research Methods and Sources of Information; (3) Nature of Social Phenomena; (4) Social Investigators; and (5) hypothesis in Social/Geographic Research and (6) Conclusion are separately dealt within a separate chapter.

Geography is considered as an art as well as science. The development of geography (from the past to the present) in the form of religion and afterwards religion was replaced by logic and a thorough reasoning and later on everything is subjected to tests, which is nearing and entering the circle of science, on the other hand "Science has assumed the sphere of the social sciences as well geography is also the field where science and scientific method and through this method of research and knowledge of unknown fields are made available, for example Columbus discovered America.

Scientific method that helps to acquire the required knowledge about the truth and other realities of life. Now, truth should be supported by experiment method and confine the truth by scientific research. For example the sun rises in the East. A knowledgeable person may easily grasp the statement without any doubt. A layman could not accept the same in a split second. We should make use of the magnetic compass and its directions, i.e., North, South, West and East and confirm the East direction, that the sun rises in the East. Through scientific observation and calculation, one could be able to accept the above statement on scientific lines and background. It is further identified that "science deals with facts and on the basis of these facts, scientific theories are formulated. The ultimate object of all this is to understand the world in which man lives. Science can impart knowledge about all these things through scientific method, which is based on observation, collection of data, classification and formulation of new theories.

Classification of Geography: 'Geography' can be classified as science with particular reference to Geomorphology, climatology and oceanography. These three sub-fields of physical geography are considered geography as a science and also the genetic classification; the land forms are ranged into mountains, hills, slopes, plateaus and plains according to MSL. The exogenic and endogenic forces create landforms into different sizes, shapes, structures and structural forms (4S) is the key to understand scientifically the science of landforms, i.e., Geomorphology and also the present is the key to understand the past. For example Rocks and Rock structure – igneous rocks, sedimentary rocks and metamorphic rocks. Keeping in view of the origin and structure of rocks and their form, final shape, size and water permeability of rock leads to know the composition helps to form the final structure of rock.

Geography is divided into three branches broadly and primarily to understand the physical process involved and evolved as physical geography and it is based on mathematics, physics and chemistry. It elaborates that mathematics and its application in fluvial cycle, (river) physics and its application to understand Aeolian Cycle, and hydrological cycle, underground water, chemistry and its application to understand the chemistry of the seas and oceans. Land and water distribution, salinity and the great soil classification for example Alluvial soils in deltas/red soils/black soils and their geographic distribution are important aspects to understand the geographic process in it. To understand climatology to the optimum possible extent, the basics of physics and chemistry are essential for the real appreciation of science of climatology. In the same way, to understand the air pollution/water pollution/land pollution and environmental pollution, one should be able to identify and recognize the quality of air/quality of water, quality of soil and pollutants and its chemical action, reaction and interaction with and within the entire process.

Natural Science: Geography is directly related to Natural Science, i.e., Botany and Zoology. The plant and animal kingdom depends upon climatic influence. Geographic distribution of animals depends upon terrain and climate, (Rainfall and temperature) and slope. This is to say that sustainability and susceptibility are the two important factors for its origin, development and distribution upon the globe.

'Geology' is the study of the earth, relating to its formation, building process and geologic phenomena in landform structure. Geomorphology is the scientific (genetic) study of landforms and their spatial distribution. This is divided into two branches. (i) Geological Geomorphology, and (2) Geographical Geomorphology. In nutshell, different geologic epochs over the time are responsible for the formation of beds, bed strength, varying stress and strain involved in the formation process, etc., whereas spatial distribution of geologic phenomena and geomorphic structure is termed as Geographical Geomorphology like, Coastal Geomorphology, Fluvial Geomorphology, and Landforms in Karst Cycle, Marine Cycle and Himalayan Geology, and Hydro-Geomorphology and so on. It deals with the natural phenomena and one could understand it in a right perspective respectively.

Geography, History, Economics, Politics, Sociology, Public Policy Science, Public Administration and Anthropology are commonly coded as "Social Sciences". These branches also deals with the social international phenomenon and have a close relation with man and society.

According to Lundberg, scientific method consists of "systematic observation, classification and interpretation of data". Scientific method may be adopted or applied for social science research equally with slight modifications, wherever it is possible. Lundberg further defines, "Social Scientists are committed to the people with the problems which they conduct for them are to be solved, it is all by judicious and systematic observation, verification, classification, and interpretation of social phenomenon. This approach in its most rigorous and successful form is broadly designated as scientific method."

L.L. Beard has designed the steps of the scientific method in his definition of the method, "Science may be defined in terms of six major processes that take place within it. These are testing, verification, definition, classification, organization and orientation, which includes prediction and application".

Karl Pearson also designed various steps of the scientific method. It follows with the following features.

[1] Careful and accurate classification of facts and observation of their correlation and sequence.
[2] The discovery of laws by the aid of the creative imagination.
[3] Self criticism and the final touch stone of equal validity for all normal constituted minds.

Scientific method must possess the characteristics like (1) Verification, (2) Definiteness, (3) Objectivity, (4) Generality, (5) Helpful in prediction and (6) System or systematized method. Social research has been defined into different forms. According to P.V. Yung, "as a scientific undertaking which by means of logical and systematized techniques aims to:

[1] Discover new facts or verify and test the old facts.
[2] Analyse their sequences, inter-relationships and casual explanations which would facilitate reliable and valid study of human behaviour.

According to W.S. Monose, "research may be defined as a method of studying problems whose solutions are to be derived partly and wholly from facts."

According to E.S. Bogardus, "Social research is the investigation of the underlying processes operative in the lives of persons who are in association."

Keeping in view of the existing definitions on social researches have the following characteristics. They are: [1] It is related to social life; [2] Social research investigates and verifies facts about the society and social interactions; [3] Social research formulates laws in regard to different social phenomenon and the principles that guide these phenomenon; and [4] Social research also tries to investigate the relationship that exists between various facts and phenomenon of the social life."

It is also identified that the motivating factors of social research are:

[1] Curiosity about the unknown;
[2] Desire to identify, recognize and know the casual or cause and effect relationship;
[3] Presentation of and appearance of novel situation; and
[4] Investigation and verification.

Finally, it is identified that, "new social situations, motivate the social research to take the social research and find out solution to the new problems and situations that have crept in" (Quotation mine please).

Types of social research: Social research typology is classified broadly into two categories. They are:

[1] Fundamental Research; and [2] Applied Research.

[1] Fundamental Research: The development and evolution of new ecological systems, new methodology of social research definitely and periodically find a place and changes have to be injected in the present research methodology. This leads to an innovative step which helps to identify, recognize and evaluate the social phenomenon in a society in different geographic regions.

[2] 'Applied research'—As stated by P.V. Young have to be examined by the investigator of the social phenomenon in any area/region.

"Gathering knowledge that could aid in betterment of human destiny is termed 'applied' or 'practical' research.

Regional planning and development studies and exercises satisfy curiosity, provide satisfaction and adds to social principles and welfare. Social research is one such topic and object. The gains of social research are: [1] Satisfies curiosity and provides additional knowledge; [2] Social welfare; [3] Development of social science; [4] Social progress; [5] Social control; and [6] Practical utility or application.

Methods and Sources of Geographic Information

Depending upon the aims, objectives and subject matter, many methods are available for collecting geographic information from different sources. Data collection, classification, framing of hypothesis, formulation of its content, geographic research not only uses a variety of methods but the approach to the problem is vivid and varied. Each method has its own merit depending upon the environmental site and situation. Geographic research methods are different from other physical and social science research. The following are important methods, that are used/adopted by geographers in geographic research from the past to the present day. They are [see Fig.3.1, schematic representation of geographic research].

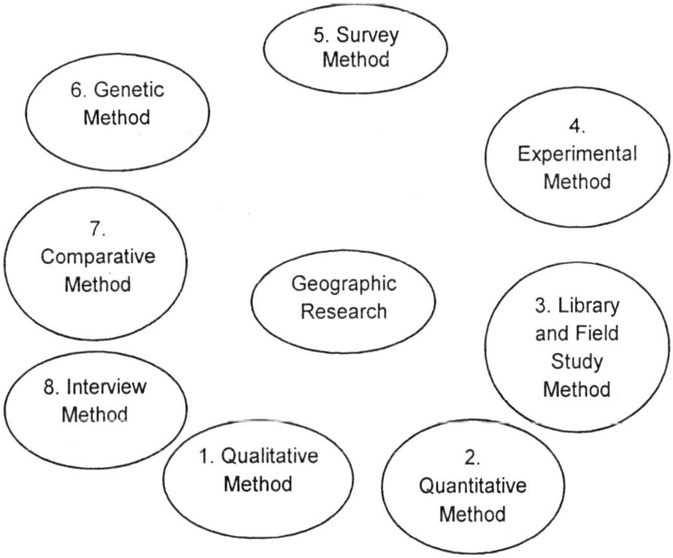

Fig. 3.1: Schematic Representation of Geographic Research

Geographic research has been classified into 8 categories:

[1] **Qualitative Method:** It is based exclusively on reason and acts are collected qualitatively. Mainly, Inductive and deductive methods of logic are involved and conclusions are drawn. Observations, description, personal study, and interview, etc., are put to a qualitative test. Historical events have a bearing upon qualitative aspects. Mainly, interview and description are of two-fold. Qualitative variables and its

description play an important role in geographic research besides social sciences.

[2] **Quantitative Method:** Statistical methods are used to an optimum level helps to derive certain valid conclusions. Statistical method helps to analyze and measure the data besides presenting facts. A social science researcher studies the family, its size and standard of living of people over a period of time.

[3] **Library and Field Study Method:** Research activity to some extent depends upon library. To complete the research activity, the researcher should be able to collect the information through this method. Much time is saved by adopting this method.

Field Study Method: Ninety per cent of geographic research and geographers depend upon this method. Quantitative methods could be applied if the researcher is able to sustain with the knowledge to accommodate. In this method, the researcher goes to the field and helps to prepare documents.

[4] **Experimental Method:** Experimental methods generally depend upon controlled conditions and environmental influence is more upon experiment. But in some experimental methods, both exogenic and endogenic influence is more and more and the results are also have its own influence. Generally, this method is carried out in laboratories. But, the geographic research particularly in physical geography, ninety per cent of research have to be done in the field only. Primary data will be collected through this method in particular to geomorphology and subfields of geomorphology are Fluvial geomorphology, glacial geomorphology, slope geomorphology and geometry Hydro-geomorphology and coastal geomorphology particularly beaches and coasts.

Collection of soil samples are done in the field and later on, they will be analysed in soil testing laboratory to identify the pedalogical patterns. Water samples are collected in the sites of different geographic sites relating to different environment and the samples are analysed qualitatively and quantitatively collection of sediments in the field are quite important to assess the rock structure and its associated phenomenon. (Beaches and Coasts). In Environmental geographic research problem, the investigator should collect the soil samples, water samples, air and its associated phenomena, sound pollution, in order to study the causative factors and the effective factors of the environment and land-form and land-use for analysis and arrive at a valid conclusion.

[5] **Survey Method:** 'Field Method' is known as external method which is away from laboratory and the primary data could be collected by the investigator/or researcher, whereas the library method is equally important and complementary method to survey method and it is to identify important characteristics with regard to the investigated/investigable problem. The scope is limited in the laboratory method whereas in the survey method, the investigator will go to the field, collect the data and from the hypothesis he draws valid conclusions and later on, he tests it.

[6] **Genetic Method:** In this method it is observed that today's politics is tomorrow's history. It means to say that historical events certainly have a bearing on political events/phenomenon or the subject matter of study. The researcher makes an attempt to identify and find correlating factor of events to the valid conclusions from his study. Origin, evolution and growth are evaluated by the researcher, while investigating the problem for research. Time, nature and environment are the important explaining events in this research in particular. This method is also known as hierarchical method, it is to say that, step by step, the events are so arranged while digging or narrating the events in a chronological way to deduce the facts and arrive at a conclusion. The unity and uniforming factor help to find various events/things to describe the body of research.

[7] **Comparative Method:** The researcher may be able to compare various facts/facets and data with one another. This method helps to draw certain conclusions. In geomorphic research, this method may be quite useful for comparing the events in the field and the recorded landforms in topo sheet may be compared through the resulted phenomena of the aerial photo interpretation. Mapping the geomorphic features/landforms helps to understand the geomorphic processes and further to understand the process which helps to arrive at a valid conclusion on scientific lines. By applying this method,

it is helpful in assessing the unity and uniformity of the unit between various things.

[8] **Interview Method:** The researcher will conduct the interview to the incumbent in order to get the right answer's to his questions serially. This type of research will be conducted particularly in social sciences, i.e., History/ Economics/Political Science, Public Administration/ Sociology/Social Work/Demography/Psychology, Geography and Philosophy. By applying this method, the researcher may be able to find the precision, reliability and validity of data. This is a new and innovative method wherein, the researcher tests the attitude/correctness of the content of the subject throughout. In this method, the researcher may be able to abstract the subject from the person both the facts and incidents that have occurred over a period of time. Time/ spatial structure could be identified and by posing more questions, relevant answers could be obtained by the researcher. Due to the application of this method, different scales and schedules could be drawn. This process helps to build social process/social phenomenon. In social science research the interview method has been occupying an important and pivotal role and assumes paramount importance. This method helps the researcher to get more touch with the subject as well the researcher may be conversant with the latest knowledge could also be obtained. This is considered as a reliable method and is quite useful in modern building of research in social sciences, particularly, geography.

Major Steps in Geographic Research

The following steps are important for adopting them in geographic research. For a successful research in geography, the researchers have to take precise and accurate steps to obtain precision and accuracy to a maximum possible extent. Certain precautions may help the researcher to obtain the accurate results in his research.

Major steps are as follows:

(1) Selection of the problem; (2) Study of research literature; (3) Preparation of bibliography; (4) Definition of the problem; (5) Analysis and classification of elements of the problem;

(6) Investigation and study of material related to problem; (7) Research design; (8) Collection of data; (9) Tabulation and systematization; (10) Analysis and interpretation of data; (11) Verification of the problem and hypothesis; and (12) Generalizations and projective part.

(1) Selection of the Problem

As we go on observing things, site and situation of the geographic environment, we may be able to select the problem. Selection of the problem always depends upon in which way we identify and select it. That is to say that the research problem which the research scholar selects should be on cent-percent should add as a new knowledge besides the existing knowledge. The selected research problem should be gainful of adding knowledge (Additional one). In social sciences, the changes should be identified, through this, knowledge may be derived. The problem should be innovative and it should be resourceful source to gain additional knowledge through this study on scientific lines.

(2) Study of Research Literature

Through intensive efforts of researcher's research from the library sources and also the rough books, the researcher may be able to identify a research problem. Review of research literature, (through this source) and the contributed additional knowledge by the previous researchers, the present researcher/investigator may be able to identify the "gap" in literature (research). Basing upon it, the research scholar certainly identifies the research problem by himself through literature survey. New thought by the researcher, the original research problem for his research is possible and plausible by the scholar. Research investigator should have to conduct a literature survey for information with regard to the problem, for a proper understanding of the problem through the series of information, the researcher may be able to collect the information at the primary level and connect the gaps by additive approach as a source of new information (knowledge) should be in a position to add new knowledge. In this way, the research scholar should be capable of identifying a research problem (will grow) from the research literature. It is said that conducting research upon research is one and only one possible way to identify a new and innovative research topic.

(3) Preparation of Bibliography

The use of library information and survey of literature is considered as a major and important step. The survey and study has a combined effect, which results bibliography to acquire the required knowledge and skills to identify a research problem and it should be gainful knowledge that provides indeed to the existing survey of literature. That is to say that the problem conceived by the researcher should be able to conceive knowledge as well to help to add new knowledge through his studies. It should be a new platform of research for the next generation too, as the researcher's contribution is one way of new thinking and still for the new research scholars they should be able to do further research, after this research scholar. This should be a connective link of studies from the past, present and also to the future generations. PP&F, i.e., "Past, present and future" are the three-way steps to the present researcher to identify and recognize it through his intensive efforts and should have an effective and interactive role which affects the present and future generations.

(4) Definition of the Problem

The researcher himself may be able to express his ideas through the identified problem for which he is able to contribute to the existing literature. The central idea of research is to identify the gap and such gap should be identified as a research problem through this, many people may be able to identify it as this is a problem and the definition should convey something as new knowledge.

(5) Analysis and Classification

The researcher should have to analyze the problem through the collected data (primary) and subsequently, if necessary, he should be able to utilize secondary data to add, deduct multiply, divide and deduce many things and finally the researcher may be able to add knowledge and communicate it to the scientists/scholars as a new body of knowledge. Here, the thumb rule is D3 and A1. Where D1 to D3 includes deletion (D1); D2 (Deduct); D3 (Deduce) and A1 – Additional knowledge.

This can be written in an equation:

RS = (D1 + D2 + D3) – A1
RS = A1 (New knowledge)

Where RS: Research Studies and A1 (New Knowledge)

In this way the detailed Knowledge acquired through his study includes the efforts of research step by step follows deletion, deduction and deducing out of the total study of material and efforts.

(6) Investigation and Study of Material

For investigating any problem, it depends on the study of the material. The problem will not be available readily or readymade one for investigation. The research student/post doctoral fellow should be able to identify the tips from the data. For example Population of India/State/District/Mandal/Village are available through Census of India Reports. It is understood that the population growth is not uniform but it gives us clue (gap in growth of population). That clue may be a step and direction towards identifying research problem as well as formulating an hypothesis.

This is to say that, a research scholar who is to undertake research studies for his M.Phil/Ph.D. may be able to identify the continuous/discontinuous way of increase in population over a period of time. Population count from 1911, 1921, 1931, 1941, 1951, 1961, 1971, 1981, 1991, 2001 and 2011 and so on (Year of Population) population figures for 1911, 1921, 1931, 1941, 1951, 1961, 1971, 1981, 1991, 2001 and 2011. The population figures for different years (10 years) may give us a clue to identify the continuous/discontinuous growth of population. This is the first step to identify by a researcher to induct/deduct the factors for identification for further studies. The investigation and study almost helps in taking future steps that are necessary for conducting scientific research in population studies by the researcher.

(7) Research Design

It is on the part of the research scholar, that he should be able to prepare the research design. In advance, the research scholar should prepare it (research design) which helps for carrying out it systematically, sequentially and scientifically. Sometimes, there are certain drawbacks in the preparation of design and this design helps to test the hypothesis, analyze data and take other necessary steps.

(8) Collection of Data

Once the researcher prepares the design, he may be able to collect the necessary data for the research problem for investigation. Collection of primary data is a primary step from the field. For example, for a geo-morphological problem, the researcher should be in the field to collect the necessary data through fieldwork/by using survey instruments such as plane table survey/Dumpy level for mapping (collecting) the landforms and connecting the same to the current site (location) and situation. This is considered as a golden step to move further and further/secondary data also should be collected from different sources and it should help for the upward process of successful research.

(9) Systematization and Tabulation

The researcher out of curiosity, he may collect more and more information but, in many cases, it helps lesser and lesser to utilize such data. The data should be categorized and systematized for further processing (tabulation) for identification and recognition. The collected data helps the researcher to categorize/systematize and tabulate because all the data is with him and he is with it. Without this, he cannot move a step further to proceed in research.

(10) Analysis and Interpretation of Data

The researcher will be in a position to classify, systematize and tabulate and he may be able to analyse and interpret. Things can be found out through good analysis and interpretation clearly by "cause and effect" factors and identify and recognize which should be eventually helpful in different possible angles for proper understanding of the subject matter and the research problem.

(11) Verification and Hypothesis

The analyses and interpretation stage certainly yields, enable the researcher to formulate certain valid hypothesis which is possible only upon verification. Next step is to test the hypothesis and then results are ready for correct assessment. Verification of hypothesis gives us good results based on scientific theory, supplemented with quantitative statistical methods, and includes parametric and non-parametric.

(12) Generalization and Projective Aspects

General principles could be arrived at if once hypothesis has been proved to be correct as a result of verification.
Finally, it is understood that scientific research and applied research may give fruitful results and also further help to contribute something new in the name of "good research".

Sources of Information for Geographic Research (S.I.F.G.R.)

The first stage of a researcher is to select a research problem and the necessary information, i.e., to collect the data relating to the research topic. The collection of data lies with the researcher and the nature of the research topic. The research scholar should have to collect the data from various sources and has to collect the data by himself in the field. Then it is called as primary data and also he can make use of the data if available, collected by others, i.e., from other sources other than by him and this is called as secondary data. Thus the sources of information for geographic research is classified into two.

(1) The primary sources of information; and
(2) Secondary sources of information.

(1) The Primary Sources of Information in Geographic Research

Under this category, the primary source of information is generally collected by the researcher by himself from the field. This will grow and comes out by the researcher from the field work. In physical geography, by making use of the instruments operation and obtains the data from the field by conducting plan table surveying for mapping the landform structures (geomorphology).

In a climatological research problem, the researcher installs the rain-gauge and takes the rainfall data of a particular station. This is turned as primary data. If the researcher wants to take temperature data, he has to take the temperature values in centigrade from the field he selects.

The rainfall data could be collected from the rain-gauge stations which was already recorded and collected by Indian Meteorological

Department, Pune. The District Collectorate/Mandal Officer of the respective Mandals have been maintaining the recorded values of temperature and rainfall (at the Mandal level). Throughout the nation, Meteorological observations relating to temperature and rainfall and the same data has been preserved by them. The researcher can make use of the same data as a secondary data for analysis by applying various statistical methods.

In Oceanographic research, the researcher should have to collect the data relating to oceans and seas, i.e., surface temperatures/salinity/aridity, etc. Water samples for analysis, the researcher should have to collect from the field.

Data relating to socio-economic and cultural variables (other than physical geographic problem), the researcher have to collect the data from the field as well through the designed question/answer (interview), study of the consumer behaviour, conversation, etc. According to Elhouse, D.N., primary data in the following terms.

> "The primary data are those which are collected for the time and are thus original in character; whereas secondary data are those, which have been collected by some other persons and passed through statistical (personnel) at least once."

(2) Secondary Source of Information (S.S.I.)

The secondary source of information have been collected by different organizations of Government of India as well as by respective State governments. For example, Census of India, New Delhi; National Remote Sensing Centre (NRSC), Hyderabad; All India Soil and Land Use, New Delhi; Indian Meteorological Department, Pune; National Sample Survey, New Delhi; National Geophysical Research Institute, Hyderabad; National Institute of Nutrition, Hyderabad; and Geological Survey of India, Calcutta, are considered as premier institutions, and they have been collecting the data from different sources and for different purposes. For every ten years period, Census of India has been collecting data relating to National Population for National Count to get final population figures throughout the country/nation, this operation has been taking place for various aspects such as physical/economic/social/cultural aspects of each State (District-wise/village-wise) data have been collected and subsequently published by Government of India, New Delhi.

National Remote Sensing Centre, this is the premier organization fully involved in survey and field work and the operation, i.e., Sensing the remote areas by different methods, i.e., taking photographs from air, thus aerial photographs are prepared and also through satellite. Satellite images are helpful to identify and understand the landform and land-use's (physical variable). Socio-economic variables relating to rural and urban settlements are also important. Settlement system could be identified and structure through the aerial photographs as well as with Survey of India (SOI), Topographical Sheet (T.S.). Aerial photo interpretation could be done by the researcher with the help of the base map as toposheet and aerial photo elements are super-imposed and traced out, as a result the changes could be identified upon the ground over a period of time (two points of time). It is collected by an organization and it is not possible by an individual. The data collected by other person (other than researcher) is called as secondary data. This is not original in the sense that it has already been collected and processed and seen by others.

Sources of Primary Data

The primary data is collected by the researcher by himself from the field that is to say that no second person will interview and interact between himself and the subject that he is studying. For this purpose he employs the following sources:

(1) Case study;
(2) Indirect verbal survey;
(3) Questionnaire Method; and
(4) Local reports.

(1) Case Study

The researcher collects the information from the field in order to collect the data both physical as well as economic, social and cultural variables. The researcher has a direct contact with the person/subject and his main objective is to collect the cent percent data to the optimum level with minimum cost factor (C.F.). The researcher may be able to get abstract information to the maximum possible extent to cover the uncovered aspects so far. In order to collect the data, the researcher should know the language and language should not be an hindrance as he has to collect the data. Social and cultural aspects of the area (location), environmental conditions and other factors relating to the area, which the researcher has to investigate for his research studies. Once the researcher is well versed with language, he may be able to conduct the field work with cent percent satisfaction.

(2) Indirect and Verbal Survey

If the researcher selects a small area for conducting his research models, he has to face difficulties (if the area is small, he has to work intensively) and if the area is large, he may not be able to conduct the field work with cent percent satisfaction of the researcher and he may not be able to contact many people because the area is larger enough to carry out the field work to an optimum level. In the present circumstances, the researcher has to depend or rely on others and he should have to make a personal study and also to obtain some important information/matters, he should take the help from others and he has to take their information also and the same will be incorporated.

(3) Questionnaire and Schedule

Geographic or socio-economic problems could be identified and recognized only through information. This information could be possible by conducting a survey through a questionnaire and structural schedule for collecting the required data which could be analysed. Social structure is a complex one to understand it structurally. It could be studied by one and only one method, i.e., by instituting it by an individual through a survey. Research cannot be successful within a day or two. It is a continuous process to intervene with individuals, institutions and intercepting causes also should be identifiable. A researcher should be able to design a schedule by himself or by a group of individuals and also to consult men with pleasing manners. This could also be in the form of a questionnaire is defined as a bank of typological questions in a structural way, which enables him to get good answers from the individuals/institutions. In essence, it is regarded as a file of subject matter of the questions. The questions should have to create inquisitiveness and curiosity from the respondent to know it in simple terms and conditions. The respondent with terrific interest puts in a series of questions to the researcher in the form of "what, why, where and when", i.e., $W_1+W_2+W_3+W_4 \rightarrow$ gives successful research. If the researcher could get the answers from the respondent, the researcher may be able to get answers for the above questions and are answered fully, it is a successful research. In the equation form it is as follows:

$$W_1 + W_2 + W_3 + W_4 \rightarrow Q_1 + Q_2$$

Where Q_1 = Qualitative questions = answers
Q_2 = Quantitative questions = answers

The creation of interest to the incumbent (who gives information) is more essential and the essence of research and its utility should be understood within no time.

1. Individual to Individual

Survey is conducted by the researcher to get good information through answers of the informer (incumbent) first Method.

2. Individual and Institutions

Survey is conducted by the researcher actively with the institution/ institutions. This is a second method.

3. Schedule Method

In this method, the researcher keeps the schedule with him/with himself. The researcher conducts the interview with the incumbent and almost all the questions are answered by the individual and the subject matter will be covered. In uncovered questions it will still remain to say that all the questions are not answered fully and necessary steps have to be taken to get full information. In this way, the research informatics will be available and the answers could be within by the individual (researcher) or sometimes generally, it is done by the incumbent (information giver). In schedule, the research investigator fills the schedule, whereas in questionnaire, the answers are within the information will be recorded by the incumbent.

4. Local Reports

Primary data is collected by the researcher, and the information provided by the respondent will be recorded simultaneously and also local reports play an important role, which forms as primary data collection. Various aspects of the research problem and the multi level information could be collected with regard to the identified research problem. Sometimes, the researcher could get the solutions through local reports, people are well versed with the information and the researcher could get it from them in a polite manner but not by compulsory method 'Social-interaction' is a key to get excellent information in research activity. It is one of the golden steps to the researcher to collect the information (data) both qualitatively and quantitatively. The information so gained from them has a greater role and importance which assumes paramount importance. It leads the researcher to step up for the neo-systematic format to get information. Thus the information forms and attains formative structure and the researcher may be able to elaborate it and it helps to a maximum possible extent. In this way, the primary data is considered to be a heart functioning and the secondary data is equally important to supplement whereever it is necessary to incorporate in the body of research and is finally considered as blood circulation and heart functioning of research body. Primary data is by all counts as heart functioning of research.

Sources of Secondary Data

It is stated earlier that the secondary data is not original one because it is collected by some individual/individuals/organizations private or autonomous organization or by the State Government or Government of India, whoever it may be.

Secondary data is classified into two forms. They are:

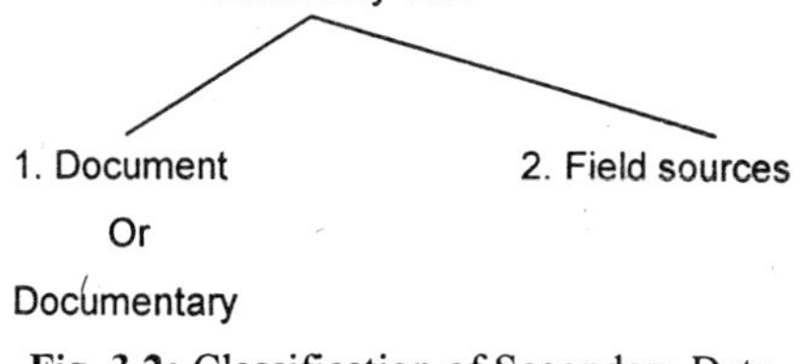

Fig. 3.2: Classification of Secondary Data

(1) Document or Documentary

The information contained in that is well documented and the incorporated information is true and the researcher can accept and adopt it faithfully and sincerely by the researcher in his research. The preserved information is passed on from generation to generation and the rich document gives us rich information. The document need not be published and sometimes, the information is left unpublished because of various reasons. We should consider this document as primary one to gain information from these published and unpublished sources. This is one of the main sources (document) to get necessary information. The researcher may have to depend upon this information as well, the researcher could put it in black and white, collections made by others can also be taken into consideration.

> "Document is either in unpublished form or published form, forms as a documentary evidence to support the researcher's argument (thesis) while presenting the valid reasons on cent percent count and no one can challenge it in any angle or on any count" (quotation is mine please).
>
> "The adequate evidence in support of the reasons while aggregating the research form firmly in a triple way, i.e., structure, process and stage and the structure of the research, structure depends upon the hypothesis and the researcher may be able to present the points cohesively" (Quotation is mine please).

Keeping in view of the above environmental site and situation, documents play a vital role in building up of the research problem to a large extent and it assumes an important role.

Kinds of documents. It is seen from the documents that the documents are classified into two, parts, i.e., 1. Published document and 2. Personal document. In this, we find, i.e., life history, personal letters and diaries (see Fig. 3.3). The field sources are presented in Fig. 3.4

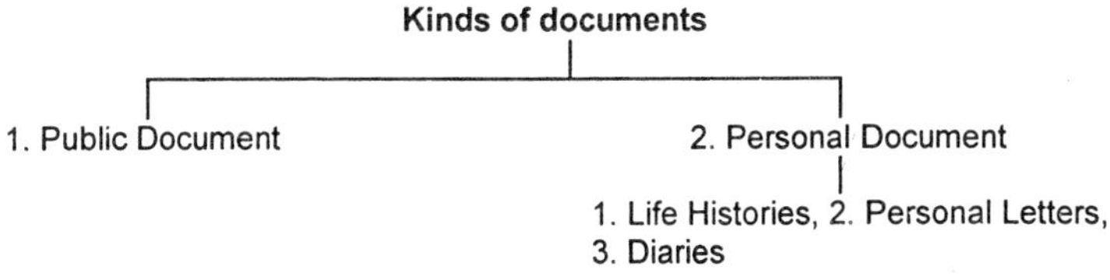

Fig. 3.3: Kinds of Documents

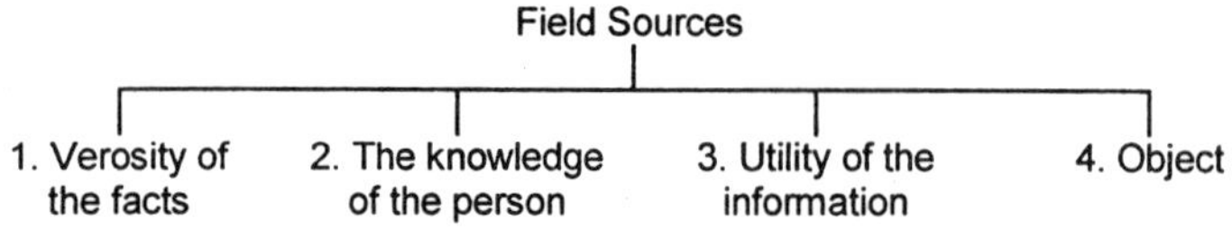

Fig. 3.4: Field Sources

Geo-Social Phenomena

'Social phenomena' may differ from place to place. So, spatial variations in social phenomena can be observed and it is termed as "time-space" element. The proper study of mankind is man and in the same way proper study of social process is called as 'social phenomenon'. In social research, the researcher/the investigator is the pioneer to control the process of research. Whereas in science, the external and internal factors are responsible for obtaining the result. The external and internal influence is being influenced on the final factor of result.

According to Curt Le Vin.

> "The social scientist should be clear that he needs all (verifiable) dynamic facts rather than symptoms and appearances and that even better, these facts are the important points of reference, like for hind and for social".

Characteristics and qualities of geo-social phenomena and Social phenomena are observed the change over space and a period of time, whereas natural phenomena is a static one and it may not be influenced by any element. Geo-social phenomenon varries

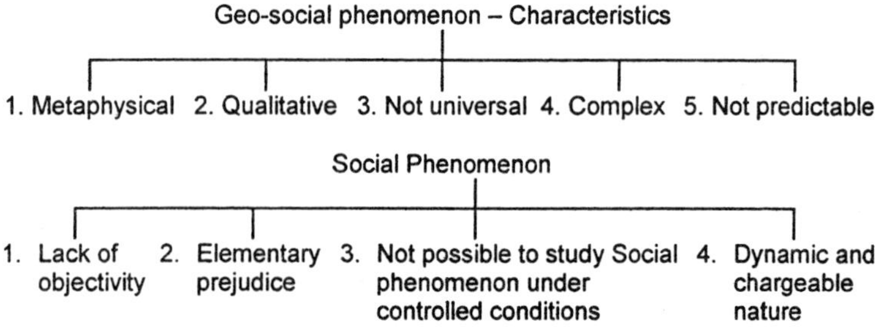

Fig. 3.5: Characteristics and Qualities of Geo-social Phenomena

from group of individuals or caste factor or environmental factor. Inmigration and outmigration are considered as two social forces which drives in and drives out, the basic phenomenon is economic factor. Economic factor influences individual behaviour and tends to migrate, where financial factors influences him to go out, this may be due to change in occupation. Sometimes social phenomenon has different characteristics. It can be seen from Diagram 3.5.

Social characteristics of an investigated problem

(1) Interest in and knowledge of the subject matter;
(2) Knowledge of research technique and interest in the research;
(3) Objectivity and contention of problem;
(4) Experience and training; and
(5) Understanding of person, place and time.

The real scientist as described by Young:

> "A scientist avoids personal and emotional interpretation of data. He is a noted debater taking issues with a sight. He maintains an open mind and tests his evidence and assumptions. He looks for the fact which is substantiated and gives theory a new meaning and vitality".

Hypothesis

Once the investigation completes the process of literature survey with regard to his research problems, he has to identify the gaps in literature survey. Such gaps should be identified by the researcher/investigator and he has to feel the essence of the felt gap and should be able to grasp it at once. At this juncture, the researcher will be entering into next phase, i.e., formulating hypothesis. It is an important phase. Once he enters into it, he should be able to formulate hypothesis. Therefore, the researcher has to identify various aspects of the research problem. By formulating a working hypothesis a considerable guidance is obtained for the other steps. The researcher gets the full idea after the completion of data collection and he may be able to start doing things in a right direction. This stage prompts the researcher to possess a considerable guidance is obtained for the other steps. This stage prompts the researcher to possess good ideas but not in a coherent way. But, the researcher wants to classify the data. But he has to form an idea and working principle based on scientific theory/ postulates/axioms in order to reach the objective. This process is known as 'hypothesis'.

According to George A. Lundberg :

> "The hypothesis is a tentative generalization, the validity of which remains to be tested. In its most elementary stage, the hypothesis may be in hunch, guess, imaginative ideas, which becomes the basis for action or investigation."

It is to say that "an hypothesis is considered to be a blue print of skeletal sketch, whereas the research investigation has to add fresh flesh and blood by incorporating the necessary inputs through them, the researcher may be able to get outputs based on the working rules, regulations and ready to put it for action leads to activate with his own ideas further and further to fulfil the objectives and find solutions to the chosen research problem" (Quotation mine please).

Further, it is a proposed generalization or justification which is subject to test in order to test its order and to test its validity besides attempting to deduce correct results out of it. According to Good and Hatt, it is, "a proposition which can be put to test to determine validity".

According to Webster, "A hypothesis is a proposition, condition or principle which is assumed, perhaps without belief, in order to draw out its logical consequences and by this method to test its validity with facts which are known or may be determined."

Keeping in view of the above, that the hypothesis assumes to be the deduction process. If this is verified and may be agreeable to the fact. "Hypothesis sometimes suggests explanation for certain facts. Such tentative explanation is suggested by something in the matter and by our previous knowledge."

Theory differs from hypothesis. Hypothesis is not a theory. The researcher/investigator may be able to connect the theoretical attributes by possible substitutes and constitutes an elaborative working principle often called as 'hypothesis'. 'Hypothesis' is a guiding principle and the preliminary aspects of objectives are inducted as subject of study and researcher has to take a shape to the intensive research problem. Generalizations are apparent both in hypothesis and theoretical constructs and the hypothesis is derived from theory itself. When the generalizations are tested by 'statistic' and found as correct, it becomes part of theory sometimes, there is only difference in form and in degree. Hypothesis is a pre-final stage of formulating a theory. According to Young, "provisional central idea which becomes the basis for fruitful investigation is human as working theory." Further, it is apt to examine, "William H. George has identified, and distinguished between theory and hypothesis described as 'Elaborate hypothesis' ".

According to Hilway (1964, p. 123):

> "From the research scholar's point of view, the hypothesis may be conveniently considered as a tentative or working assumption, and the theory as the surviving or final hypothesis which is most defensibly supported by all the evidence, But since knowledge arrived at through the scientific method is subject to revision in the light of new data, a theory is in only one sense always a working assumption, so that the conventional distinction between hypothesis and theory (on the basis of increasing adequacy of evidence and hence of greater certainty) is a relative one. For all practical purposes, the duality of the term hypothesis need not concern the research scholar except to serve as a salutary reminder that no conclusion reached by the scientific method is necessarily a fixed and final truth."

> "The hypothesis is powerful tool in research process to achieve dependable knowledge. It helps the researcher to relate theory to observation and observation to theory".

Hypothesis are formulated only as the suggested solution to the problem, which the objective that the ensuing study may lead either to its rejection or to its retention. They involved in the study and suggest methodological procedures that are to be employed" (Lokesh Kaul).

The Role of Hypothesis in Research

The role of hypothesis assumes (utmost) paramount importance and those who aims to make predictions about outcome. In scientific research the researcher expects positive results from his experiments. A social science researcher wants to investigate the urban history or about a nation, the life of a man, or an event seeks to facts to determine the status/position and he may not have a base for making prediction of result. To put it in nutshell, that, "a hypothesis, may not be required in such a fact-finding studies. According to Hillway that, "when fact finding alone is the aim of the study, a hypothesis may not be required". It helps to interpret the facts to generalization. A political geographer/a scholar of a political science makes a study of the results of parliamentary elections/opinion pole, the factored data he collects will definitely proves to be useful only if he is able to draw some valid generalizations out of it. In all possible ways, a hypothesis is definitely recommended for all major/minor studies in order to explain observed facts, situation, or human behaviour, it serves as a guideline in research process. A researcher may waste his time, money and human energy in collecting the empirical data without making his policies and then find that he cannot present facts lucidly and identifies the relevant relationships between physical and economic variables. Impact of physical factors on economy have its own role on many counts and variations upon outputs (input-output analysis relating to physical variables and *vice versa* as in the absence of a good hypothesis to guide him. The importance of hypothesis is stated in four points as given in Fig. 3.6.

1. Hypothesis
   2. Extension of Knowledge Trends to Tentative Explanation (facts and phenomena) (in the case of area studies)
   3. Tested (statistical)
   4. Validity leads to make concrete laws

2. Situation Event Unconfirmed Facts — logic+intelligence guesses+about Unknown Hypothesis

3. Hypothesis
   Direction to research
   Selecting sample
   Research procedure
   Application of statistical technique
   Data Analysis
   Delimit the study

Fig. 3.6: The Importance of Hypothesis

In the above case, for example, if a researcher is interested in studying the effect of rainfall on agricultural productivity that is to say that relationship of physical variables on agricultural output/ production. He hypothesizes: "There is a negative relationship between rainfall on Agricultural Production". It is crystal clear from the above statement of hypothesis that the researcher would conduct an experimental method most possibly (statistical method) in which he compared the rainfall and agricultural production (crops) to the rainfed crops and irrigated crops. The significance of the difference between the mean rainfall and mean productivity of the two situations could be analysed by using t-test or analysis of variance technique.

4. It provides the basis for final reporting of the conclusions of the study. It finds very convenient to test each hypothesis and state the conclusions that have relationship to each. Keeping in view of the above, the researcher can prepare a good research report.

Hypothesis Formulation

Three important factors helps the researcher to formulate good/best hypothesis:

Case 1

Researcher's background knowledge. A researcher may deduce hypothesis inductively after making theoretical practical observations of behaviour, noticing trends or probable relationships.

New knowledge, New discoveries, and new inventions may help the researcher to formulate a good hypothesis. Based on enriched experience and good academic background, people may be able to formulate good hypothesis.

Case 2

Versatility of intellect hypothesis are mainly derived from inductive as well as deductive reasoning from a theory.

Case 3

Analogy and other practices lead the researcher to many clues which he finds useful, in the formulation of hypotheses and for finding solutions to problem. For example: the dullness of a school-going boy and his academic achievement (Age is an important factor while formulating the hypothesis). Genetic factor sometimes play an important role in detecting the roots of illness. (Present candidate : Father/Mother/Grand-Father/Grand-Mother) and their academic achievement and background.

The interaction of a researcher with academic people, conversation and consultations with experts colleagues from different fields of specializations makes the researcher and helps him in formulating the best hypothesis.

Criteria in Hypothesis

1. Criteria is an important aspect and the hypothesis should be clearly and precisely documented. Clear cut statement of hypothesis contains technical language and definition of terms should be understood.
2. Hypothesis should be testable and hypothesis predicts the research outcome of the selected study of the area. They must be in relation to explain variable that are subjected to measurement. A hypothesis, should define the variables operationally by stating the operation/procedures that are necessary to measure them.
3. Hypothesis should state the expected relationship between variables. A good hypothesis should have to state exclusively an expected relationship between the variables.

Case 1

For "Increasing distance leads to a decreasing migration". In the above statement of hypothesis that "distance" is a powerful tool to delimit the migration process and it is time in the case of a poor men though he possesses good skills through his engineering/ medicine.

Case 2

"Intensive efforts of a candidate makes him to reach the said goal".

In the above statement of hypothesis is that a candidate has to reach the goal, a candidate has to put in intensive efforts (A positive statement). Even though the candidate put in intensive efforts, he could not reach the goal. (A negative statement). (Result is zero). Keeping in view of the intensive coaching for an EAMCET, candidates may prove to be excellent in their attainment (positive statement). Without an intensive coaching, candidate may not attain good position (positive statement). With intensive coaching (sometimes) no attainment of good rank (negative effect). Without any coaching, the candidate has attained good rank (positive effect). HO: Coaching has no effect in academic rank. (This should be experimental and tested to a possible extent) to accept or reject it.

A research hypothesis can take into three forms. They are (1) declarative form; (2) the null form or (3) question form. Keeping in view of the above yields – If a researcher makes a positive statement about the outcome of the study the hypothesis takes the declarative form. For example the hypothesis. "The rank holder (intensive coaching) in EAMCET is significantly higher than that of the *less rank holder* (no coaching at all). The researcher makes a prediction based on his theoretical formulations of what should happen, if the explanations of the rank attainment, he has given in his theory are correct.

In agreement with the Null form, the investigator makes a statement that no relationship exists. He makes the hypothesis, "there is no significant difference between the academic achievement of trained rank holder and that of untrained rank holder", is an example of null hypothesis. The null hypothesis can better the statistically, they are of the termed as statistical hypothesis. They are also called the testing hypothesis when declarative hypothesis are tested statistically by converting them into null form. It is apparent in educational and related research.

> "The null hypothesis challenges the assertion of a declarative hypothesis and denies altogether, it says that even where it seems to hold good it is due to make chance" (Lokesh Kaul, p. 103).However, inferentially the case of the declarative hypothesis becomes quite strong. The less plausible the null hypothesis the use plausible the declarative one. Haymen opined that, null hypothesis should not be used as an alternative to the declarative hypothesis but should be used in combination with it for statistical purposes". However, it

must be noted by the researcher that the question form is less powerful than declarative or null form as a tool for securing valid information."

Testing of the Hypothesis

Hypothesis are possible explanations which account for the factor, events or conditions that the researcher attempts to understand, after they are formulated according to the criteria discussed, they are subjected to empirical as well as logical testing.

1. Some hypothesis are simple and they can be tested directly. In most situations, however, they are complex and, therefore, cannot be tested directly. They have to be tested in terms of their deduced consequences.

   Indirect method of handling research problems involve intricate and complex procedures. Intellectual and disciplined effort is needed for the deduction of consequences. In this way, the researcher does not test the hypothesis but tests the deduced consequences of the hypothesis cause all the deduced consequences, after testing, comes out to be true, the hypothesis is confirmed. If some of the consequences are true and certain other untrue, then the hypothesis will have to be examined afresh.

   The necessary conditions for confirmation are:

   (a) All factual evidence collected through tests or other means should correspond with the deduced consequences.

   (b) The test situation or data collecting tool should take into account that all factors and conditions that are suggested by the consequences.

   (c) The consequences are logically deduced from the hypothesis.

2. The absence of conflict with the other satisfactory proved generalizations leads to support the correctness of an hypothesis.
3. A hypothesis is also confirmed to be correct if the predictions made on the basis proved to be successful.

Further, it is important for the researcher to formulate hypothesis before data are collected. This is necessary for an objective and unbiased study.

REFERENCE

Lockesh Kaul, "Hypothesis" from H.K. Misra: *Research Methodology*, pp. 95-106.

4

Design of Geographic Research – Objectivity – Geographic Survey

"Design of geographic research" or "research design' play an important role in geo-scientific research. The next step of formulation of the hypothesis and it has to be tested (subject to tests) for identifying the orderly form or not. This is termed as empirical test of hypothesis. For this, the investigator has to prepare a research design, through this design of research, he may be able to test the validity of the hypothesis (in agreement with the hypothesis) on the basis of the analysis of data and outcome (tested). Design is nothing but adoption of type and technique of geographic research which is most suited for this research and study, investigation and the research the problem, proper or apt material has to be collected for the selected field of study. Data collection determines the study and it is to be carried out according to the structural design. In this way, design is, therefore, becomes the integral base and basis upon which we proceed further and further care in studying the problem after formulating the hypothesis.

According to Russel Ackoff, "Design is the process of making decisions before a situation arises in which the decision has to be carried out. It is a process of deliberate anticipation directed towards bringing and unexpected situation under control."

According to E.A. Suchman, "A research design is not a highly specific plan to be followed without direction, but rather a series of guide ports to keep one headed in the right direction."

According to Jaheda, Deutish and Cook, "A research design is the arrangement of conditions for collection and analysis of data in a manner that aims to combined relevance to the research purpose with economy procedures".

Keeping in view of the above, the research design is articulated as. *"searching by the researcher a research design in order to collect the necessary data relevant to the research problem is the primary duty of a researcher and the research design articulated by the researcher himself leads to the successful completion of research problem* – and it is to state that,

"Research design is considered to be a framework in which the researcher has to collect the data according to its stipulation to avoid unnecessary confusion at a later date in terms of time and money without any deviation to its mechanism in a true perspective (static and dynamic factors)". [Quotation mine please]

Formulate design or exploratory design is the essence of research design which supports the argument and it depends, totally on the methods of collection of data and analytical tools and techniques, through which analysis takes place. Design occupies an important place in the field of geographic/geo-science research. The role of research design "the structural design should be able to schedule/ questionnaire in addition to the unanswered questions, should have to be answered by the researcher (guess, trial and error) in agreement with the formulated hypothesis" (Quotation mine please).

Keeping in view of the above site and situation, 'environment' is the key factor to control the entire built up of the landform (aggradations and degradation process) particularly the fluvial/ Aeolian cycles (site and situation depends upon 'environment'. Design is very much determined by the problem, its nature and nurtured process and data. It has a vital role in the field of geographic research and the role may be stated as follows.

(1) Definition of the problem and measurement procedures;
(2) A research has to distinguish the inclusive and exclusive factors of research;
(3) Diagnostic study; and
(4) Adoption of new design (research) for new elements of design in order to study a particular problem.

Types of Research Design

(1) Exploratory design, (2) Descriptive or Diagnostic design, and (3) Experimental design.

(1) Exploratory or formulate design contains social significance, the problem should have practical aspect, combination of theory or theoretical combination of theory or theoretical facts. Jahoda and Cook have pointed out, "Exploratory research is necessary to obtain the experience which will be helpful in formulating relevant hypothesis for more definite investigation."

"Investigating problem should have practical utility and direct to the public welfare subfield to area, people and economy" (Quotation mine please). It is to stay that many answers should be able to give as far as possible over a period of time (time bound and objective bound) research.

Exploratory or formulate design helps to study the following aspects such as (1) Information about the environment, (2) Presentation of important problems, (3) Study of the unfamiliar and unknown field, (4) Theoretical base, and (5) Presentation of uncertain problems for study in research.

Goode and Hatt (p. 92)

Further problems in research design: Some explanation is needed in the following paragraphs. Such as (1) Delimitation of the factors to studies; and (2) Further variations of research design. Goode and Hatt have opined that, "it is easier to answer a good question than a poor one. The emphasis is laid on seeing the problem in theoretical and conceptual terms. They are:

(i) Studying the research literature on the subject;
(ii) Analyzing the problem with colleagues and teachers, particularly with those who have worked on the problem;
(iii) Attempting to locate unpublished material or projected research on the subject;
(iv) Attempting to carryout formally the steps in conceptual specification;
(v) Developing a sharp detail the hypothesis, including its connections with physical/social/economic theory; and
(vi) Actual contact, where possible, with the phenomena being studied

A pilot study, connects to aiming at and to as certain which are the salient field problems and the important selected variables. The augmentation of depth interviews may be quite useful as well as all types of observation and even the application of various hypothesis to see whether they seem to fit (good, fit) the situation. Generally, a pilot study will be limited to a few cases which are chosen strategically as possible.

It is a matter of practicability to the investigator while investigating the problem. We cannot investigate all the interesting aspects of the problem and at some point we must arbitrarily limit its size. However, by being aware of this limitation, we are at least leave open the possibility of later interpreting our results within our research which does deal with the variables have ignored (Goode and Hatt, p. 93).

(1) One of the variables to be delimited controlled or taken account of, is the observer himself.
(2) Each observer has his personal a cultural frame of perception which determines in part which he sees to deal with this problem, requires standardized techniques for both observing and recording the data we need.
(3) The knowledge on the part of the individuals or groups that they are being observed may change their behaviour.
(4) When the social researcher is using existing data, such as agency records, census figures, school files, etc., his own inference may be minimal, but he must be aware that some one made the observations originally and that earlier effect must be weighted.
(5) Both the questionnaire and the interview are modes of interaction between observer and observed, and the role of the former is a part of the system of variables in the problem itself.

Keeping in view of the above situation, "it is observed that research design which clearly states the problem controls the relevant variables and selects an appropriate logic of demonstration is difficult of achievement but worthy of great effort". (Quotation mine please)

Objectivity in Research

'The subjective study' and 'the objective study' are the two important factors in any type of research. Subjective study is that in which the likes and dislikes, feelings and attitudes of the investigator's influence on the study, while in objective study, the investigation remains, passive spectator. He does not view the subject of his study or the social phenomena from his subjective point of view or through glasses object by his personal feelings/attitudes, etc. The objective study presents a dependable study. Objective study is the first connection of scientific studies. According to Carr – objectivity of truth means that the phenomenon world is a reality independent of beliefs, hopes or fears, of any individual, all which we find out not by intuition and speculation but by actual observation.

Social problems do get influenced by the subjectivity of the investigator. The difficulties may be identified and studied under the following heads.

(1) Emotional effects, (2) Complexity of the subject matter, (3) Lack of uniformity, (4) Qualitative nature of the subject matter, and (5) External pressures.

How to achieve objectivity is a major problem to the investigator in any type of research. The following are certain measures through which objectivity may be achieved. The investigator has to make every possible step/effort, to avoid prejudices and bias and for achieving objectivity, the following measures are to be adopted:

(1) Use of empirical methods and statistical measures;
(2) Standardization of terms and concepts;
(3) Use of schedule and questionnaire – In order to achieve objectivity in the collection of data, the questionnaire and schedule are quite useful and helpful. 'Standardization' is important and powerful tool in good research;
(4) Group Research;
(5) Use of experimental method;
(6) Use of Random Sampling Method;
(7) Mechanical devices; and
(8) Use of inter-disciplinary techniques. This method is very much in vogue.

To conclude, it may be said that if proper techniques and methods are employed, objectivity can be achieved in the field of geographic research as well. The element of objectivity has to be achieved if the results of geographic research are to be depended on modern methods of research available in the chosen branch/inter-disciplinary research methods are to be adopted.

5

Sampling Method and Experimental Method

An attempt is made in this chapter to present the analytical framework of sampling method and experimental method. We have to take into consideration the questions like (1) what is sampling method, Definition of sampling (few definitions), basis of sampling—importance of sampling—Advantages of sampling—Disadvantages of sampling. Probability and sampling (Goode and Hatt, p. 208-231); (2) Types of sampling method—(i) Random sampling; (ii) Stratified sampling method; (iii) Purposive sampling; (iv) Multi stage sampling method. (3) Selection of sample characteristics of a good unit; size and reliability of sample – sampling method in the present day study.

Part 2, deals with experimental method – Meaning – definition – types of experiment – experiment and its planning – methods of selecting control groups – merits and demerits of experimental method – geography laboratory and the validity of results.

The investigator/researcher should have to decide whether the entire population is to be covered in the total research for data collection or a particular part or a segment of population is to be selected as representative of the entire population. It is a big and herculean task for moving forward in his research activity. The first method, when the entire (total) population is taken into account it is termed as 'census method', whereas the other method is that, "when a small group/segment (part) is taken into account as representative of the whole (universe), it is called 'sampling method'.

Definition of Sampling

The method of sampling is defined by various eminent scholars, are as follows:

According to (1) Frank Yates, "The term sample should be reserved for a set of unit or portion of aggregate and material which has been selected in the belief that it will be representative of the whole aggregate".

(2) William J. George and Paul K. Hatt, "A sample as the name implies is smaller representative of a large whole".

(3) According to P.V. Young, "A statistical method is animation picture or core section of the entire group or aggregate from which the sample" is taken (Scientific Social Survey and Research).

Keeping in view of the above, the method of sampling is defined as, "some of the number/numbers are disaggregated picked up one, from the total aggregated items, which effectively represents the whole population and the uncovered population characteristics should have to be covered by the picked up aggregate" (Quotation Mine Please). It is understood that the sample should represent the entire population in all aspects and in all respects do possess statistical validity for testing and so on. The percentage which is known as group/'sample' has a representative character and the basis of sampling is as follows.

(1) Homogeneity Phenomenon and Heterogeneity Phenomena

Social/geographic phenomena is a complex one and compounded with different characteristics. In simple terms it is quite difficult to define homogeneity/heterogeneity. In socio-economic technology, we may find/investigate social heterogeneity/economic inequality, which exists due to many reasons 'Homogeneity' is defined as "similar characteristics (built up) with pedagogical characteristics. For example, a delta, is defined as homogeneity in terms of soil character as sand, silt and clay – a plain area from the mean sea level with less undulations and angular make up. Finally, it looks like a triangular shaped landform and the outcome is formed with the help of river's final action. It is the resultant landform of rivers final activity. It is said that physical homogeneity could be identified with the built up of socio-economic heterogeneity. Variations in socio-economic levels of development, i.e., social heterogeneity/economic inequality may be identified in a delta and one can take up for critical study. In spite of complexity and compounded diversity/diversifying characteristics, the group has certain fundamental traits that are common among all. For example, Krishna delta/Godavary delta or Mahanadi delta. Heterogeneous characteristics are to be identified among a set of social group such as University teachers/College teachers. For example, we have to carry out research amongst the university or college teachers, we can select a group of people from a particular university/college. It is evident (true) that no two teachers are similar. It is to say that dissimilarity could be identified among a set of homogeneous group, the character (college teacher) of heterogeneity is seen in their behaviour/attitude and so on, but they have common traits in them, therefore, the study that is based on a group, shall be called true of the study of the whole mass or the entire population of university/college teachers.

(2) Possibility of Representative Selection

It is seen and verified that if we select a number of units or people from a mass, simply on random basis. In this way, the group selected is the representative of whole mass. This assumption is based on "Law of Statistical Regularity".

(3) Absolute Accuracy Not Possible and Essential

In the undertaken research project by a researcher, undertaken the mass study, it is not possible to get the exact and cent per cent accurate results. In the field of geographic/socio-economic research we have to depend upon "averages", interpolated and extrapolated figures of population for analysis.

Sampling method, it is not the absolute accuracy but significantly identified as accurate, which is needed, in particular, large scale study and observations. On the basis of the sample and its analysis could be used to make certain generalizations, statistically valid, when they are subjected to tests and verifications.

Importance of Sampling

'Census Method' cannot be adopted because of the financial implications besides time constraints. So, people do resort to 'Sampling Method' to avoid financial implications and social complications. In any research activity, while conducting sample study, people have to co-operate and co-ordinate with the investigator and their team. Otherwise, this activity could not be conducted on sound statistical lines. The required data will be acquired as per the specifications. Representative units are reliable and dependable and this method saves much of the time, energy and financial sources.

(1) Study of Large Area

Study of large area could be possible only through sampling method. If the study area is large, more time will be consumed but the entire 'picture' (outcome) could be understood.

(2) Study of Small Area

Intensive efforts should have to put in by the researcher because, the area is small and the researcher have to put in more efforts in order to collect the information which helps for understanding its 'totality' (universe) complete survey helps more in a small size unit. It is possible to study intensively the units and their phenomena.

(3) Scrutiny of Variable Information

In other methods, it is difficult to test the accuracy and veracity of the results.

(4) Facilitates the Research in collecting the Information as per Schedule/Questionnaire

On the whole, we have to confine ourselves to the selected items, in such a way that the researcher is on an advantageous position for easy and quick collection of data/information.

(5) Attainment of Sufficient Results

In this method it is possible to get large number of results in a short span of time.

Advantages of Sampling Method

Six advantages are visible and possible, such as, (1) Economy of time, (2) Economy of resources, (3) Detailed study, (4) Accuracy of results, (5) Administrative convenience, and (6) Avoiding difficulties, while conducting 'census method' of survey.

The disadvantages of sampling are as follows. There are five listed disadvantages in this method. They are as given below.

(1) Possibilities of bias and prejudices, (2) Difficulties in selective representative sample/lack of representation, (3) Need for specific and specialized knowledge, (4) Not easy to adhere to samples, and (5) Impossibilities of sample.

Characteristics of a Good Sample

A good sample is that which fulfils the correct objectives of the research. A sample in order to be good should have the following characteristics. (1) Representative character, (2) Sufficient number of samples, (3) Selection free from bias and prejudices, (4) Continuity to subject matter and means, (5) Use of common knowledge, and (6) Use of past and other practical experience.

Probability and Sampling

> "Since an important part of sampling depends upon probability theory; it is first necessary to digress somewhat by discussing the nature and application of probability theory. One or two ways of looking at probability in so far as scientific method is concerned."

Probability refers to the likelihood that a given statement is a time statement. This conception of probability relates to the amount of knowledge lying behind a statement whose 'probable truth' is being evaluated. The conception of probability which is fundamental to the sampling process.

The second conception of probability is essentially a mathematical one. It holds that probability express the frequency of the occurrence of a given event, in any series which could produce either occurrence or non-occurrence. For example: If a coin is tossed into the air it has an equal chance of turning up heads or tails (Since it has only two sides). If heads are designated as 'P' and tails are 'Q' then, the probability of securing heads may be stated as

$$P = \frac{1}{2}$$

The probability of securing tails is also 1/2. The generalized form of this statement is

$$P = \frac{1}{2}N$$

Where N = the possible number of events. Thus the number of heads expressed in 10 tosses of a coin would be $\frac{1}{2}(10)$ or 5 heads.

The probability of securing more than one of a series of mutually exclusive events is secured by adding the probabilities for each. Thus if the chances of throwing a 6 on a die are equal to 1/6 and the chances of throwing a 1 are also 1/6, then the chances of throwing either a1 or a 6 are equal to 1/6 + 1/6 or 1/3.

If these two events are considered as one, however, the calculation is different. Thus, in two throws of a die (or one throw of two dice) the probabilities are different if we wish both a1 and a 6. What, for example, there are the chances of throwing a 7 in a 'crap game'? The accompanying table shows the calculation. The probabilities of each of the mutually dependent pairs is equal to 1/36. Since all are mutually exclusive ways of making 7, the sum of the probability values yields the probability of throwing away 7, which is 6/36, or 1/6.

Table 5.1: Ways of Making 7

Die 1	*Probability*	*Die 2*	*Probability*	*Combination Probability*
6	1/6	1	1/6	1/6 x 1/6 = 1/36
1	1/6	6	1/6	1/6 x 1/6 = 1/36
5	1/6	2	1/6	1/6 x 1/6 = 1/36
2	1/6	5	1/6	1/6 x 1/6 = 1/36
4	1/6	3	1/6	1/6 x 1/6 = 1/36
3	1/6	4	1/6	1/6 x 1/6 = 1/36

It should be noted in these examples that an assumption of equiv probability is made between alternative possible events such as throwing heads or throwing non-heads. What is there to justify the assumption that heads are just as likely, no more and no less, to appear as are tails? Or that each of a dice is exactly as likely as any other side to occur? One answer of item is 'chance'. But then chance must be defined. Hagood says, "chance is generally described as the result of a multiplicity of simultaneously operating independent factors".

Another way of putting this is that we know of no reason to expect any are more frequently than the other and thus we make the assumption of "equal probability". The question of heads and tails can be used for illustration. It is recalled that the probability of head on the toss of one coin is ½ probability each of the possible combinations when two coins are tossed? Since, as in the case of the dice, these related results, the probabilities are multiplied. Thus the probabilities are:

$$2 \text{ heads} = \tfrac{1}{2} \times \tfrac{1}{2} = 0.25$$

$$2 \text{ tails} = \tfrac{1}{2} \times \tfrac{1}{2} = 0.25$$

$$1 \text{ head and } 1 \text{ tail} = \tfrac{1}{2} \times \tfrac{1}{2} = 0.25 \quad \text{These being the same,}$$

$$1 \text{ tail and } 1 \text{ head} = \tfrac{1}{2} \times \tfrac{1}{2} = 0.25 \quad \text{They are added} = 0.50x$$

X is regarding the order, that is whether a head or a tail precedes the other.

It can, therefore be said that in a considerable number of tosses of two coins, the types of results will be in the proportion of 2 heads (0.25) + 1 head and 1 tail (0.50) + 2 tails (0.25) = 1.00 (the full range of possibilities). Now it happens that there is a general mathematical expression for this relationship which can express any number of possibilities rather than the three simple possibilities given in the example. This mathematical expression is known as the expansion of the binomial and takes the general form (P+Q)n, where 'n' is the number of events. In the case illustrated above, this formulae would yield the following:

$$(P+Q)^2 = P^2 + 2PQ + Q^2$$

In which P^2 refers to 2 heads, PQ to 1 head and 1 tail and Q^2 to 2 tails. The frequencies, expressed as the coefficients of these terms, are (1 for P^2, 2 for PQ) and 1 for Q^2. These yield the same proportion as the earlier example. Namely, 0.25, 0.50 and 0.25 respectively. When the binomial is expanded to that 'n' represents a large number of events, the number of combinations increases and the frequencies of each form a perfectly smooth systematical curve. This is the familiar "probability curve" or as it is sometimes called the curve of normal distribution. "This curve is fundamental in probability statistics."

Types of Sampling Methods

Various methods are being used for selecting and drawing the samples. They are (1) Random sampling, (2) Stratified sampling method, (3) Purposive sampling, (4) Multi-stage sampling method.

1. Random Sampling

In this method all the units are given equal importance. Every unit has the possibility of being included in the sampling. It is also known as "proportionate sampling" because of entire population is according to pattern,

> "Random sampling is the form applied when the method of selection assumes each individual or element in universe an equal chance of being chosen".

While applying random sampling method, it should be done with the help of certain methods, they are:

(1) Lottery method, (2) Tippets numbers method, (3) Grid system of random sampling.

(1) **Lottery Method:** Writing the numbers of names each unit and putting them in a container. They are thoroughly mixed and contain numbers are picked up from the container and those which are picked up, are taken up for sampling.

(2) **Tippets Numbers Method:** It is evolved by L.H.C. Tippet who constructed a list of 10,400 four digit numbers written at random on every page. From them one can choose 50 samples are to be selected for study, out of the total number of 500, then we can open any page of Tippet numbers and select first 50, that are below 500 and take them up for study.

(3) **Grid System:** This is used for an area to pick up samples. In this method, a map of entire area is drawn and should be divided into *equal* squares and some of the squares are selected at random. Example. Equal sizes independent, accessible replacement should not take place.

2. Stratified Sampling Method

The universe or the entire population is divided into a number of groups/states. This method deals with 'strata' and also it is called stratified sampling method. Certain numbers of each item are taken from each group at random. It should be selected with a definite purpose. There should be a perfect homogeneity in different units of strata/structure.

3. Purposive Sampling

In this method, the researcher selects the unit for certain purpose for study. Adolph defines it as, "purposive selection denotes the method of selecting a number of groups of units in such a way that selected groups together yield as quickly as possible the same averages or population negative totality with respect of those characteristics which are already a matter of statistical knowledge."

Based upon statistical knowledge, purposive sampling denotes that the universe is divided into groups of strata/structure representing each special characteristic presenting the topographical difference with socio-economic difference and thus so selected for study which yields definitely the entire universal character in total or each segment gives aggregately representing characteristics of a universe" (Quotation mine please).

4. Multi Stage Sampling Method

Experimental Method: Scientific research based on experiment which helps to identify and recognize the 'cause' and 'effect' factors and the investigation may be able to test the hypothesis. Natural science (Botany and Zoology), Physical Science (Physics and Chemistry) and Human or Social Sciences have its own experimental methods to identify and check the 'process' which involves the causative factor as well as effect factor. For example Biodiversity. Biodiversity is caused due to the influence of climate (rainfall, temperature and soil factor) and change in climate leads to a picture of Biodiversity, whether it may be apparent in a geographic unit or regions (many units). In the branch of Zoology, the influence of geographic conditions on animal and animal kingdom leads to further formation of zoo-geographical regions on a large scale. For example, marine environment on animals and related lives.

If the researcher is keen to observe and identify the Newtonian physics or thermodynamic laws in Chemistry, as these two may be applied to the real/true situation of an urban phenomena (such as in urban geography/urban economics/urban sociology). The analytical form is stated in its quotes.

"The increase in distance leads to a decrease in (consumer) or urban facilities/utilities" (Quotation mine please) "The internal energy of a thermodynamic system depends on it but not the path followed by it" (Thermodynamics). If we logically deduce from it, we frame in such way that, "The efficiency of a person depends upon him but not the path followed by him" (Hypothesis, mine please). In the case of human or social sciences, the scientific experiment takes another way. If we want to observe the migration pattern, it takes in this form.

Case 1: "Heterogeneous factors in a homogeneous plain causes (creates) anxiety to human living", and

Case 2: "Population modality (migration) leads to human brain drain (internally) and external forces have a telling effect on internal economy".

Case 3: Influence of extraneous factor on economy and people, the inflow of money (external source) leads to a hike in land values. (Cause → Migration) and effect → land values. Inflow of money (external source) and the internal object is land value is effected in a particular region/area.

American dollar makes an indelible mark of price on land values in the essential area (Migration prone area – people).

Finally, we conclude in logical terms that, "external factor have its own influence on the existing internal factors such as land and in particular agricultural land".

"Proper investment (capital) leads to create internal employment otherwise triggering of unemployment and creates new situation".

"Keeping in view of the above statements or events made the researcher to make such statements upon the investigation he has to verify the structure/process/stage of the economy and suggest suitable measures to curb and ultimately direct the investor to make investment in a proper way, which benefits society at large" (The cause and the effect are the two factors leads to two types of human mobility causes results to outmigration and inmigration. Finally, "it effects the very primary process of 'Human Migration' and creates/helps to internal stability and security due to external money lending situation. Otherwise the breakdown process in agriculture may be marked and visualized".

The researchers have to conduct experiments as per the design of the experiment. Experiment may be able to add knowledge and create efficiency to the existing human resources (scientist). The framing of scientific laws is possible only if the experiments are conducted on scientific lines and helps to create creativity among the subjects. Now, the author/researcher would like to shift upon various experimental situations which warrants to take up at present. Keeping in view of the past and present situation in order to safeguard them from dangerous situation, one has to put in efforts to curb stagnation and stalemate situation and fruitful efforts to create employment as well as to utilize internal sources to a maximum possible (optimum) level to achieve the desired and targeted results over a period of time, i.e., planned development for optimum employability of people and helps to arrest migration process.

Experiment Method in Geomorphic Research

Primarily, the researcher has to identify and recognize the land structure and later on, examine the existing internal and external forces. At that time stage, he has to enlist them in a sequential form. Before proceeding to field work, the researcher should have preliminary information relating to 'location specific' and 'human specific' factors of the selected area to conduct experiment (field work) with surveying instruments to map the landforms into existing phenomena [Horizontal/vertical/spatial form (length and width)] including slope (slope analysis) angles; etc. Observation and measurement and the two important factors and the third factor is to identify the location with reference to basics (area map administrative/topo-sheet) and neatly draw them in the field note book).

Periodical surveying should be conducted (time-space) specifically on the earlier locations in order to obtain and deduce the changes that have caused over the space (place) in a timeframe by superimposing them upon the previous one. It gives (X_1-X_0) and later on (X_2-X_1) Drawing of Beach profiles. The volume of change could be identified and the interpretation could take place by the researcher easily, specifically and spatially.

Even the socio-economic forces also have its impact on the economy and society. In order to identify the factors at the micro level, the researcher has to collect data from the field (villages) and compare the same to the macro level units to calculate the ratios, percentages and proportion to what extent? 'why' and 'where' and cause and effect factors also could be identified with valid reasons. Spatial and temporal variations could be identified as causative factors and affect factors lead to changes in socio-economic and cultural sectors.

In socio-economic survey, the researcher has to conduct experiment on one group of population and the other group is said to be controlling group. In order to maintain the similarity, certain steps have to be taken in selection of control group. The steps are as follows:

(a) Precision Matching,
(b) Matching frequency distribution, and
(c) Randomization.

In this type of survey or in this type of experiment, "the control group or groups play an important role in this type of experiment. John Franck remarked rightly, "Ideally, the control group should be matched with experimental group of all compounding factors matching on the basis of easily measured variables such as size of group, age of the subject, etc., will do good unless these variables are systematically related to the hypothesis to be tested".

Merits and Demerits of Experimental Method

Merits

(1) *Scientific Method*: This is very valuable method in all Natural/ Earth Science besides social sciences.
(2) *Determination of Cause and Effect relationship*: Through this method, it can be done quite accurately, because the researcher can control the conditions under which a particular thing occurs.
(3) Useful in testing the hypothesis

Demerits

(1) *Difficulties of Setting*: Experimental method can be very accurate for natural sciences because they are carried out in the laboratory and these settings can easily be created. It is very difficult to set a social experiment in its proper setting.
(2) Difficulties to get co-operation and co-ordination in conducting the social experiment.

Role of Geography Laboratory and Validity of Results

Geography Laboratory: The primary as well as secondary data have to be processed. Once the researcher collects the information from the field, he has to check them according to layout of the design and procedure. If it is genuinely conducted by the researcher, he may be able to get good response in the name of 'results'. For collecting the primary information from the field, the investigation should invariably know the operational technique of the instrument, otherwise there is no use at all.

(1) Surveying Instruments

(1) Plane table with stand,
(2) Prismatic Compass,
(3) Drawing sheet,
(4) Pencil,
(5) Eraser,
(6) Clips,
(7) Box compass,
(8) Metallic tape or cloth tape,
(9) Camera (loaded with film),
(10) Toposheet, and
(11) Base map.

These items help him to conduct plane table survey and mapping could be done. Drawing of profiles in a beach if it is on the sea coast – coastal geographical research particularly coastal geomorphology.

(2) Secondary Data should be made available in Geography Laboratory

(1) Topo-sheet,
(2) Base map,
(3) Aerial photos,
(4) Pocket aero master,
(5) Computer: To prepare map on computer and digital photography,
(6) Tracing table,
(7) Calculators,
(8) I.M.D. Weather maps, and
(9) Drawing instruments.

Climatic Data

(1) Rain gauge,
(2) Thermometer. Six's Max and Min thermometer, and
(3) Hygrometer – and also investigator should collect samples of air, water and soil samples in the field relating to urban/rural field work.

Oceangraphic Data

Water samples, sand samples (Samples collection bags) sieves to conduct size and shape analysis. Cartographic instruments should be made available in order to prepare maps in series for comparing them with the base map, to detect changes 'spatially'.

A Well Equipped Research Library

The research library must contain references and standard books on the subject and also periodicals, journals and magazines that are helpful for further study of the problems.

Rooms for conducting Conferences, Seminars and Symposia

Field charts, maps, thematic maps, administrative maps, physical maps, geomorphic maps, climatic maps and depicting ocean floor, etc., should be made available.

(1) A centre for geographic information/cartography.
(2) Trained staff for conducting field work.
(3) Technical know-how.
(4) Processing of data.

The centre itself helps to investigate and carryout research. Since it has a good deal of equipment and researches, it makes the research possible in a shorter period of time. It is concluded that geography research laboratory should be well equipped and enriched with the above instruments besides the maps, charts, computer and a well trained cartographer/photographer to undertake geographic research in any branch or on any problem seldom selected by the researcher/ investigator.

6

Data Collection and Inputs in Geographic Research

An attempt is made in this chapter exclusively the data collection methods, it is to say that the collected data is treated as inputs in geographic research. In the earlier chapters, we have dealt the major branches of geography (Chapter 2); Nature of geographic research methods and sources of information and Nature of geographic phenomena, Hypothesis in geographic research are described in Chapter 3; and in Chapter 4, Design of geographic research – its objectivity and geographic survey is presented. Sampling methods and its typology is presented in Chapter 5 inclusive of experimental method and sampling method and role of geography laboratory and validity of results.

In the present chapter, a humble attempt is made to present various research methods of data collection internally and externally both published (secondary) and unpublished (primary data) besides field based collected data which is considered as necessary research inputs for further proceeding with the problem with in a time-space perspective activity.

The data collection methods are:

(1) Observation Method, (2) Participant Observation Method, (3) Controlled Observation Method, and (4) Organization of the field observation – Merits and demerits;
(2) Questionnaire Method – types – merits and demerits;
(3) Schedule Method – types – characteristics of good schedule – framing of schedule – schedule and question – pilot study merits and demerits;

(4) Case Study Method – importance;
(5) Library Method – Its utility and limitation;
(6) Interview Method – types – technique of interview – role of interview – validity of interview – merit and demerits – techniques of internship; and
(7) Documentary method of study-types; and
(8) Technological inputs.

1. Observation Method

The primary function of an investigation/researcher is to observe the field, men and material that are existing. The researcher should have an unquenchable thrust for knowledge in the subject as well as field work. Observation is the beginning of research problem in the direction of collecting the material and the researcher should become investigator as well as examiner of various things in the field. To start with men, material and animal and be able to manage to grasp information relating to area-specific and location specific. His target should be to collect the maximum information/informatics within a shortest period of time in the field. Observation is considered as scientific and the researcher should be more scientific enough to observe the ongoing process. Though, the researcher is a new person to the people but he should be more friendly to manage the people and situation. It is further to state that he is the best Manager of human resources as well as the event manager of the event management. In this method, the researcher/investigator should have to manage the time element also as far as possible and will be able to establish good relationship with the people whether administration/lowest people of the administration. Researcher and respondent (interview) relationship should be established. The researcher should be able to collect the information/data (valid) from the field without asking anybody later on he has to establish rapport with the people in order to obtain the necessary information relating to the research problem which is well connected with the field/within the field and of the field. The objective of the researcher should be fulfilled, only if the researcher is enriched enough to manage the situation with his skills.

According to Goode and Hatt, "Science begins with observation, and most ultimately return to observation for its final validation. The social scientist must then train himself to observe carefully" (Method of Social Research, W.J. Goode and Paul K. Hatt).

Observation is dependable and perfect method of social investigation and not only established and popular but mostly reliable for achieving/getting/gaining knowledge of social phenomena. Observation is the classic method/scientific enquiry and the dictionary meaning of observation stems into this form, i.e., "Accurate watching and nothing of phenomena as they occur in nature with regard to cause and effect or material relation."

Nature of Observation

The observation should be unparallel. The eyes, ears and hands should be used for observing the men and things, while conversation, the researcher should be able to grasp what people are giving information and he should be able to record the events that are available in the field. It is stated that the researcher should be more curious and cautious enough to get information, why curious – he has to get the information; and why cautious? He should be cautious enough to adjudge things in the field and also he has to collect the primary data from the field.

(1) *Scientific Observation and Deep Study*: The investigator should have to investigate the material based on scientific observation and deep study. Keenness leads to deep study and collection follows. Observation should be planned, recorded systematically and it should be subjected to further periodical checks, crosschecks and controls for the further goal of the quality of research of the research problem.

(2) It is selective and purposeful collection of data from the field.

Types of Observations

The types of observations are stated below:

1. Simple or non-controlled observation,
2. Controlled observation,
3. Participant observation, and
4. Non-participant observation.

1. Simple Observation

In this, the researcher observes the things and record the observation (environment-ecology-development). No attempt is made to utilize instruments to achieve precision or check for accuracy. In an uncontrolled observation, through his participation, he is able to acquire the required information from the people and also in the field.

2. Controlled Observation

This is called as systematic observation based on logic and reasoning. The researcher should be seasoned enough to reason and completely depend upon people (sample) control over the observation/observer. According to Goode and Hatt, "Systematic observation have control over the observation and observer". Further the researcher should have planning of observation, schedules, and mechanical appliances, etc., leads to proper functioning of research by himself. Detailed notes, group/team observation and the information from people will help the researcher to a large extent in his investigation.

3. Participant Observation

Participant observation is typologically an uncontrolled observation. According to Young, "the participant observer, generally using non-controlled observation, lives or otherwise shares in the life of the group which is studying".

Merits

(1) The researcher can acquire wide range of information;
(2) The researcher is able to have a direct contact with the measures of the group of respondents; and
(3) He may be able to study the real behaviour and acquires time information from the respondents as per the size of the sample.

Demerits

(1) It requires longer time and resources are utilized much and resources may be utilized in a short span of time; and
(2) Due to emotional attitude of the researcher, sometimes hampers the objective of the research.

4. Non-Participant Observation

This is also a method of uncontrolled or natural method of observation.

Merits

(1) Objective observation is possible;
(2) Dependable information could be acquired;
(3) Influence has no role and the researcher is able to acquire information; and
(4) Able to maintain status as a researcher.

Organization (Research)

(1) Time, place and people are to be observed intensively and extensively;
(2) In the field, the researcher have plans to meet eminent people to have an interaction with them for the required data and so on.

It is concluded that, "observation is not only a popular method but also it yields dependable and correct results". In spite of the limitations, it is a scientific and class method and Moser stated that "Observation can fairly be called classic method of scientific enquiry".

2. Questionnaire Method

This is a method structured with various types of question, relating to research topic to collect the answers from the respondents by the researcher. This method is in practice from the earlier times to today/ todate, particularly in the areas of social science research. Questionnaire method is based on a series of questions to be asked to respondent by the investigator. That is why this is called as questionnaire method. The schedule method also contains questions but the nature of questions may be designed to a timeframe not to create non-cooperative attitude of the respondent with the researcher. The principal aim is based on the objective of the research is unit and it is intact.

According to Bigardus, "A questionnaire is a list of questions to a number of persons for them to answer. It secures standardized results that can be tabulated and treated statistically." Further, "in general, the questionnaire refers to a device for securing answers to questions by using a form which the respondent files by himself" (Goode and Hatt).

"It is to state that, in order to collect the necessary information by the researcher from the respondent throughout the study area as shown to achieve success in collecting reliable and dependable informatics (it is a device)". (Quotation mine please).

According to purpose it is designed as (1) structural questionnaire; (2) Non-structural questionnaire – closed questionnaire, open questionnaire, pictorial questionnaire and mixed questionnaires. Young defines it as, "the closed form of questionnaire usually contain itemized answers to the questions being asked".

Characteristics of Good Questionnaire

According to P.V. Young, "A questionnaire is answered at a distance with the assistance or supervisor of a field worker. It is, therefore, essential to keep constantly in mind the factors which are most likely to stimulate correct replies".

The questions should be analytical, clarity, limitations and no room for emotionality and emphasis properly besides physical form of questions and then it is termed as pool questionnaire. The advantages of this method ranges to less cost (expenditure), possible to cover the large area with greater reliability and validity of information subject to free from external influence. Disadvantages of this method ranges to difficult situation to frame questions to study all intelligence levels, and there is no scope for detailed study, lacking personal contact.

3. Schedule Method

According to Webster a schedule is, "a found list, a catalogue or inventory".

"Schedule is the name usually applied to a set of questions by the investigator where questions are asked and filled in by an interview in a face to situation with another para" (Goode and Hatt). The purpose is to collect the requisite data. In this method the researcher puts certain questions and the respondent furnishes certain answers and the researcher (interview) records them as the answers are furnished. It facilitates and help us when tabulation and analysis data is taken up.

Accurate communication and accurate response is designed in the schedule, it is turned as good schedule without contiguity. The content of the schedule should be with (1) an introductory put, (2) Main schedule, and (3) instructions.

4. Case Study Method

Generally in social sciences, this case study method is adopted and it is more reliable and responsive study. Case study method is a study, which ranges to location specific in its terms and designed to cover the uncovered aspects by which creativity could be instilled.

According to Goode and Hatt, "it is a way organizing social data so as to preserve the unitary character or the social object being studied."

"It is a study relating to locality (area) belongs to population of an area who acts in multiple roles and the researcher's research has the limitation of time and space with specific objectivity subject to the exposition of human behaviour in a spatial form and suggest the lines of development to the governance finally". (Quotation mine please).

"This is a qualitative study of the researcher for an administrative region/regions within a geographic limit and puts the quantitative form of suggestions (inputs) for the alleviation of people and area from below, through the inputs such as infrastructural plan to augment it by the governance, if it is accepted for generating employment and other aspects and so on". (Quotation mine please).

Merits of Case Study Method

(1) Study of social unit by the researcher is possible and eventually picks up the problems and offer solutions.
(2) Contact could be developed during the study period and take up the suggestions/opinions and prepare them finally to take a materialized form.
(3) All the aspects such as people, and area with specific problem could be studied by the researcher.
(4) It is a useful method for the study of interaction of variables and hypothesis could be tested through the available statistical methods.

A Case Study – Procedure

In a case study, normally the following form/stages are followed to complete the study successfully:

(i) Statement of the problem – (a) selection of cases, (b) types of units, (c) number of cases, and (d) scope of analysis.
(ii) Description of the course of event.
(iii) Determinants or factor or factors – (1) General factors, (2) Special factors, and (3) Additional sources.

Additional Sources: (1) Books and Magazines, (2) Literature survey, (3) Identifying the gaps in literature, and (4) Government documents, communications. "For obtaining qualitative data from the field, the researcher/investigator should have to collect the data from the field for primary data and also he should collect the secondary data from different sources and the analytical points which are more valid should be well connected to find solutions either by the people or by the governance to solve them in a specific period or a part of it over a period of time." (Quotation mine please).

Supplementation by projective methods are to be prepared by the researcher/investigator. This case study method is a superior method and the researcher selects it in order to find and offer solutions to projected series of problems. By all its merits, it is a valid method for modern research too !

5. Library Method : A Spatial Perspective

Geographic research and library method: Both external and internal factors often play an important role which decide the quality of the research topic as well as its methodology which the researcher selects to carry out his research. Geographic research is considerably both external (field work oriented) and internal based (library work oriented) and Geography of environment should have to be blended with qualitative aspects to alleviate the results of the researcher which may find a place both internally (within the country) and externally (outside the country).

'Library' is a connective link to the researcher as well as to the field on which he carries out the research work. The investigator should be able to use the library as a perennial source of information to enable him to equip with the event or events and 'Library' is defined as the place for aggregation of encyclopedias, reports, text books, recent arrivals, journals, communications and committee report and so on. It is aptly put it in a disaggregated way according to catalogue number, call number and so on by author-wise, subject-wise and title-wise are arranged in racks. In recent times, the e-libraries are developed spatially. This method helps the researcher in multiple ways and it guides the researcher properly and the supervising research guides him scientifically and systematically to reach the objective of completing the work as per time schedule. The researcher should be able to select books and other material, make proper use, prepare bibliography and annotated bibliographies, etc.

The multiple roles are as follows:

(1) Selecting the subject matter; (2) Choosing the methodology; (3) Selection of the libraries; (4) Selecting the relevant material from the vast available material in the library; (5) Making use of the selected material; and (6) Ways to make study more useful through the library method

Do's by the Researcher

(1) The collected material from the library should categorically be numbered, dated and significantly superscribe, "research material and got typed/Xerox and get it corrected by himself as far as possible.
(2) The senior researcher's help also could be taken whenever and wherever it is necessary, if he feels.
(3) Research material when it is used in the body of a research paper, honestly, he has to put them in quotes and also write it in footnote number and also may be included in bibliography at the end of chapter. (4) latest trends in research should be identified by himself, it is possible if the research takes up his research both intensively and extensively without any time constraint.

It is observed that geographic research in its true terms it is 'spatial' and special in its location, structure, process and stages, keeping in view of the above spatial structure, its possibilities and plausible conditions will decide the ongoing research. Geographic research is considered to be field oriented as well as laboratory oriented and the 'Library' is an important source of informatics which provide the maximum information to the researcher whereas the 'field' will provide maximum information relating to research and many efforts are mostly responsible and helps for conducting qualitative and quantitative research not only in geographical research alone but also in all social sciences.

The library inputs are considered as necessary research inputs (data collection) for social science research and encyclopedias, gazetteers, pamphlets, journals, periodical regional studies, seminar volumes, conference numbers (India and abroad), case studies form as secondary data and the primary data forms own field work of a researcher. Both primary data and secondary data, combinedly have the effect which should be the final effort of researcher to collect the collected data with possible techniques, the analysis have to be carried out and the final output leads to good results in any research activity.

With regard to the secondary data is concerned, the library information forms as input for geographical research that could be conducted successfully by a candidate if he makes use of the following material keeping in view of the current informatics by different methods.

Journals are published periodically by many institutes/institutions and departments of the concerned subjects of the universities of higher learning. But, the inter disciplinary research attracts many scholars to diverge and converge their views into a concrete form to obtain good results in their research from Anthropology to Zoology, the geographer will have a connecting link with Anthropology as Anthropogeography and Zoology with Geography, which innovative branch termed as connects to Zoo-Geography 'Geographic research' has close links to many physical science, life sciences and social sciences.

Data inputs (library method) will form and ranges from different subjects like centrifugal forces and centripetal forces, which leads to results to an excellent research.

(1) Geography of Rural Development

The researchers should consult the journals entitled:

(1) *Journal of Rural Development*, published by the National Institute of Rural Development, Rajendranagar, Hyderabad, which is available at Centre for Economic and Social Studies, Begumpet, Hyderabad. Some more Journals are also

available at CESS, Begumpet, Hyderabad for consultation by the Social Science Researcher (information is as follows).

(2) *Nagarlok/Urban Afairs* – Quarterly – Published by The Indian Institute of Public Administration, New Delhi – 110 002.

(3) *Urban India* – A Journal of the National Institute of Urban Affairs, India Habitat Centre, New Delhi.

(4) *The Deccan Geographer*, Pune.

(5) *Demographic India* – The Indian Association for the study of Population/Institute of Economic Growth, Delhi University Campus, Delhi – 110 007.

(6) *Indian Journal of Social Development* – An International Journal Serial Publications, New Delhi – India.

(7) *The Indian Journal of Economics* – The Department of Economics and Commerce, University of Allahabad – 211002 (India).

(8) Journal of Indian School of Political Economy, *Arthabodh* – Pune (India)

(9) *Journal of Health Management*, Sage Publication, New Delhi.

(10) Economic and Political Weekly, Mumbai.

(11) *Indian Journal of Regional Science*, Kolkata – 700 091.

(12) *Indian Journal of Political Science Association*, Jaipur (Rajasthan), India.

(13) *Indian Journal of Public Administration*, published by Indian Institute of Public Administration, New Delhi.

(14) *Rajagiri Journal of Social Development*, published by Rajagiri College of Social Sciences, Kochi, Kerala, India.

(15) *Review of Development and Change*, Madras Institute of Development Studies, Madras

(16) *Environment and Industrializations Mapping, Enumeration and Surveying Informal Settlements and Studies*, Sage Publications, New Delhi.

(17) *Geospatial – Today*, Hyderabad, Focus on Surveying and Mapping India's First GIS Journal.

(18) *Journal of Social and Economic Development*, Published by Institute for Social and Economic Change, Bangalore.

(19) *Indian Journal of Agricultural Economics*, Published at Mumbai.

Also available at Osmania University, Hyderabad.

(20) *Geographical Review of India*, Kolkata, West Bengal.
(21) *National Geographic Journal*, BHU, Varanasi, U.P.
(22) *Geographer*, Aligarh Muslim University, Aligarh, U.P.
(23) *Annals*, National Association of Geography, India, New Delhi.
(24) *Indian Geographical Journal*, Chennai, India.
(25) *Transactions*, Pune.
(26) *Journal of Geography*, by Allahabad University, Allahabad (Uttar Pradesh).
(27) *Eastern Geographer*, Department of Geography, Bhubaneswar.

Abroad Journals are also available at Centre for Economic and Social Studies, Begumpet, Hyderabad.

(1) *National Geographic*, Washington D.C., U.S.A.
(2) *Population Studies* – A Journal of Demography, London School of Economics, London.
(3) *Population and Development Review*, New York, U.S.A.
(4) *Rural* 21, The International Journal of Rural Development, *www.rural21.com*, Frankfurt, Germany.
(5) *Environment and Development Economics*, New York. The Beiger Institute of Ecological Economics.
(6) *Asia-Pacific Population Journal*. UNESCO and SOC – Commission for Asia and Pacific, Published by U.N. Publications, Thailand.
(7) *World Development*, Canada.
(8) *The Journal of Development Studies* at Manchester, U.K.
(9) *Annals* [Annals of the Association of American Geographers], Washington D.C., U.S.A.
(10) *Ekistics*

6. Interview Method

This method is a systematized one, based on mutual contacts of people. This is a connective process or link between the researcher and the interviewee. This method helps cultivate connections at large and finally leads to interaction of the two persons, and exchange of ideas and it is an innovative method of all times. A few definitions of 'interview' are given below so as to understand the process behind its technique and technology. According to P.V. Young, "the interview may be regarded as a systematic method by which one person enters more or less imaginatively into the inner life of another who is generally a comparative stranger to him."

According to Goode and Hatt, "Neither Reliability nor depth can be achieved, however, unless it is kept clearly in mind that interview is fundamentally a process of social interaction." P.V. Young further stated that interview as, "the interview is a technique of field work which is used to watch the behaviour of an individual or individuals to recent statements, to observe the concrete results of social or group interaction. It is, therefore, a social process which usually involves interaction between two persons".

Interview is stated as, "a process of interaction of two persons of unknown to know each of them and have an interactive voice reactor (IAVOR), enabling to connect them and concrete ideas may help to develop, and it ensures to secure extensive information in an intensive way to fulfil the researcher's objective of acquiring the required data" (Quotation mine please).

Merits

(1) It helps to formulate the hypothesis through interview if it is face to face (direct interview by the researcher to a new personnel)

(2) It helps to collect information about qualitative facts, which facilitates for further improving the ideas that are exerted or emerged out due to this process

(3) It helps to locate – specifically the problem prone areas whether it is a physical problem or socio-economic or cultural problem which helps for development/development process indeed, and also it helps to estimate the psychology of a person too !

Method of Interview

(1) Direct contact method; (2) Indirect contact method; (3) Through telephone/cell phone/online/S.M.S. and e-mail. Modern techniques of interview may help the researcher to get the desired level within a shortest period of time. The level may differ from process to process of interview; and (4) Balanced and timely questions could be asked

so as to cut short the unnecessary matters which should not circumvent the object and the subject.

The last method is historical, comparative, deductive and inductive method, in social science as well as in geographical science. The data collection and the major inputs for geographic research in major branches of geography are stated in chapter 7, into five stages such as 7.1 Physical Geography, 7.2 Economic Geography includes Agricultural and Industrial Geography, 7.3. Population Geography, 7.4. Settlement Geography or Geography of Settlements, and 7.5. Regional Geography, Regional Development, Rural Development, Regional Disparities, Regional Resources and Variations in the Levels of Development." The methodology of research is particularly examined and suggested the alternative methodology for future research keeping in vicw of thc existing and suggest future research methodology in geographic and allied methods to adopt for achieving higher level of results (outputs) in research spatially and globally if not locally" (quotation mine please).

7

Research Inputs (Data Collection, Processing and Results) in Major Branches of Geography – The Ongoing Process : An Overview

In chapter 2, the major branches of Geography have been presented and the changing and ever changing role of geography should be understood. Data collection (as inputs) in major branches of Geography (Conceptual trend) and methodology have been identified and selective variables in geographic research are presented in this chapter. Major branches of Geography are presented systematically and diagrammatically (Fig. 7.1).

Physical geography is divided into three subfields. They are:

(1) Geomorphology, (2) Climatology, and (3) Oceanography (Fig. 7.2).

The major studies in geomorphology are earth crust, landforms and their transformation; major landforms—cycle of mountain development and mountain as an ecosystem, plateaus typology—plain typology, coastal plain, destruction plains, glaciated plains, desert plains, penne plains; different forms of plains—karst plains, depositional plains, alluvial plains, drift plains; till—1. Rivers, 2. Underground and surface water, the moving ice, wind, waves and shorelines, hydrosphere, lakes and swamps; soils and vegetation are intensively dealt over an extensive geographic area and the research work is going on scientific lines (Fig. 7.3).

In climatology, the major work ranges to the atmosphere, atmospheric pressures, winds, variable winds, moisture in the

6. Zoogeography (Geography of Animals) 5. Biogeography (Plant Geography)	7. Regional Geography Regional Planning Regional Development Regional Resources
4. Settlement Geography (1) Rural Settlement Geography (2) Urban Geography	8. Anthropogeography 9. Political geography
3. Population Geography or Geography of Population	10. Historical geography 11. Mathematical geography
2. Economic Geography (1) Agricultural Geography (2) Industrial Geography	12. Physico-chemical geography
1. Physical Geography (1) Geomorphology (2) Climatology (3) Oceanography (4) Environmental Impact Assessment (EIA) (5) Environmental Degradation	

Fig. 7.1: Major Branches of Geography

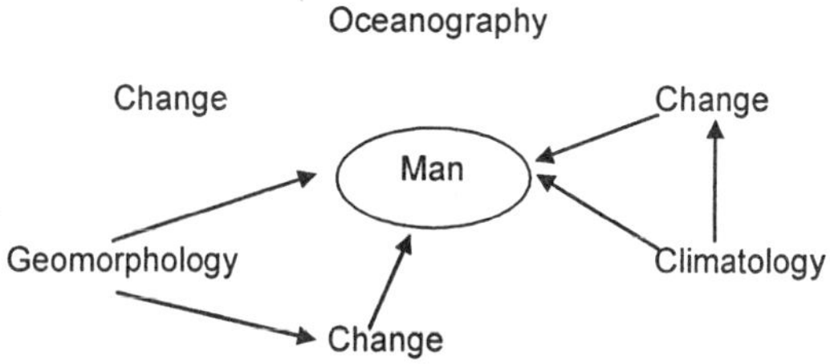

Fig. 7.2: External and Internal Forces and their Influence leaves an Indelible Mark on Man and Space (Human-Spatial Environment)

atmosphere, climate and man, climate and topography, climate and weathering, climate and denudation, climate and vegetation and the chief factors of climatic control, such are the aspects dealt in many researchers in India and Oceanographic researchers are quite promising field for the other branches related to marine geography, marine geology and marine geophysics.

Plan of the study is divided in this chapter are as follows:

7.1 Physical Geography
7.2 Economic Geography

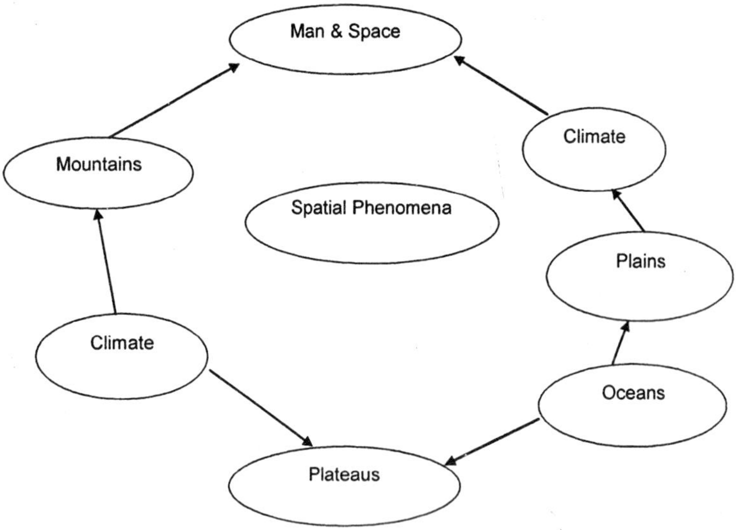

Fig. 7.3: Spatial Informatics – Applied Geo-Informatics

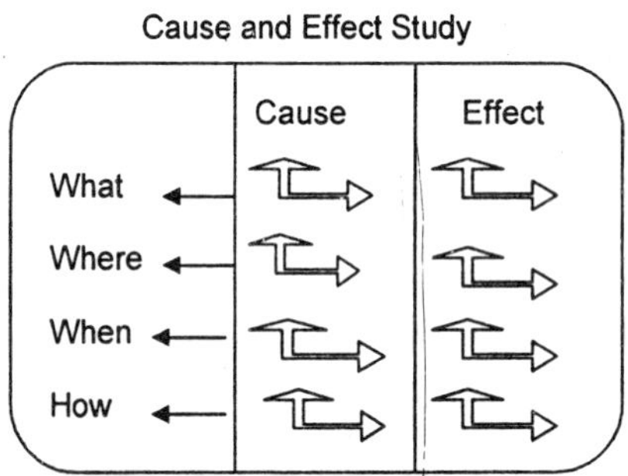

Fig. 7.4: Cause and Effect Study

7.3 Population Geography
7.4 Settlement Geography (Urban and Rural)
7.5 Regional Studies
7.6 Role of Technology in Geographic Research

7.1. DATA INPUTS FOR GEOGRAPHICAL RESEARCH IN PHYSICAL GEOGRAPHY

(1) Current trends in Geomorphology, Climatology, Oceanography

and Pedology indicators, describing development of fluvial, glacial, coastal and Aeolian landscapes particularly the nature of time in geomorphology and nurtured features upon space.

(2) Climatology majorly covering atmosphere, winds, solar radiation, earth energy budget, air pressure, atmospheric moisture including the explanation of the local, regional/ national level climate.

(3) Concentration on soil and other related issues bordering and cutting across other disciplines.

(4) Variables relating to an account of biodiversity even at the regional level and covers the problems of environmental pollution.

Land-Use Capability Classification – Micro Morphological Units

> "Land-use capability classification of the micro morphometric regions must be considered for agricultural development of an area. In land management programme, geomorphic unit like river basins is a viable unit as compared to the administrative unit as it is basically an agro-climatic unit".[1]

Geo-Environment of a River Basin

Generally, the rivers flow from west to east direction (Indian River System). The attitude of the river basin varies from highest point to the lowest point (H_1 to L_1) in metres, (miles). Undulating plain, land with laterite, upper tertiary and lower quaternary alluvium is the order and the altitude varies from higher to lower level (in metres). Swamp lands are also the prominent features of the confluence zone.

Particularly, climate condition of the basin one—hot and dry continental type. The average annual rainfall is about 1200 to 1400 mm.

The major morphometric divisions have to be postulated mainly on the basis of the contour height. Generally, the divisions may take into four forms: (1) Marginal plateau fringe (150 m.), (2) Undulating plain (120-60 m), (3) Gently sloped plain (60-50 m), (4) Lower plain land (50-30 m), and (5) Flood plain (Less than 30 m).

Generally the land units have diverse topographical, soil, irrigation and erosion patterns.

Step 1: Identification of morphometric units as it may appear in the field and have to be recognized.

Step 2: Land, water, soil, use (Productivity features) and modified factors should be identified.

Step 3: Land capability classes have to be indicated in a map and also these sub-classes reflect the degree of some major conservational problems and limitations.

E = Erosion Limitation
W = Wetness Limitation

Step 4: Uniformly sloping high and low flood plain along the river side is the land of productive soil, suitable for paddy and other crops

Step 5: Land of gently sloping riverside upland

Step 6: Parametric indices of the factors determining the land capability:

(1) Land → (a) Relief, (b) Slope, (c) Drainage.

(2) Water → (a) Rainfall, (b) Irrigation, (c) Water potentiality.

(3) Soil → (a) Texture, (b) Nutrient Status.

(4) Use → (a) Arable, (b) Forest, Scrub and grasses, (c) Wastes.

(5) Modifying factors → (a) Erosion/acidity/salinity, (b) Flood/Drought/almost immune frost affected.

Morphometric Analysis

Methodology:

Step 1: This study should be based on the integrated use of remote sensing and GIS techniques along with the supportive information collected from Survey of India toposheets (Scale 1 : 50,000).

Step 2: The demarcation of watershed and preparation of drainage map have to be done with the help of Survey of India sheets and later, this map have to be updated by adding the relevant information extracted from digital data of IRS-P6, LISS-III geo coded (FCC) Product. The GIS techniques using Arc view ver.3.2 have to be applied for mapping and computing different aspects of drainage basin such as length of stream, length of basin, areal and relief aspects. Prioritization rating of micro-watersheds have to be carried out by calculating compound parameters values based on linear, area and relief properties of sub-watersheds.

Preliminarily, study relates to understand morphometric parameters such as stream order, bifurcation ratio, stream length and stream length ratio and length of over land flow, etc., have to be taken into consideration for detailed analysis.

A Case Study

Application of multi temporal satellite data for the study of water resource problems in Udaipur basin.

Methodology: "The study is undertaken with the help of Survey of India topo sheets of scale 1:25,000 and 1: 50,000 surveyed in 1989-60, 1989-72 and 1973 respectively.

(2) Satellite images of varying resolutions for different years, viz., land sat MSS data of 3rd March, 1975 with 80 metres resolution, IRS data of LISS 1 of 18th February, 1996 with 72.5 metres resolution, IRS data of LISS III of 8th February, 2004 with 23.5 metres resolution and IRS data of LISS IV 7th June, 2004 with 5.6 metres resolution have been used.

(3) Geological Survey of India, Map 1997 depicting litho logical formation of the region has also been used. Ground truths have been verified through field survey.

(4) Secondary data pertaining to rainfall, irrigation, land-use, population lake water storage capacity, forest cover, etc., have also been acquired from various Government Institutions and department of Geography UCSSH, M.L.Sukhadia University, Udaipur.

For this study three broad methods have been used. Visual interpretation of Remotely sensed data, mathematical analysis and field survey for verification of ground truths. The satellite data have been geo referenced. Digital imaging processing techniques have been used for the demarcation of cropped area for 1975, 1996 and 2004. Catchment boundaries have been overlaid.

(5) Geological Map has been reserved for lithological composition of the entire basin and a composite picture has been arrived at for analysis lineaments, drainage, settlements and roads."

Morphometric Analysis

Methodology

(1) This study is based on the integrated use of remote sensing and GIS techniques along with the supportive information collected from Survey of India (SOI) toposheet to delineate the river basin (Scale 1: 50,000).
(2) The demarcation of watershed and preparation of drainage map are to be done with the help of Survey of India sheets.
(3) The drainage map is to be updated by adding the relevant information extracted from digital data of I.R.S.-P6, LISS-III geo coded (FCC) product. The GIS technique using Ave view ver.3.2 is to be applied for mapping and computing different aspects of drainage basin such as length of the stream, length of basin, areal and relief aspects.
(4) Calculate compound parameters values based on linear, areal and relief properties of sub-watersheds.

Data Inputs for Geographical Research such as Environmental Issues

The land cover changes severely affects the character of the soil, water, air and forest resources as shown below:

Resource	*Degradation/Depletion*	*Casual Agricultural and Production Activity*
Land	Soil erosion, micronutrient, deficiency, alkalinity/salinity, water logging	Frequent tillage, non-leguminous mono culture, non use of organic manure over irrigation and flooding of fields, canal irrigations improper drainage
Water	Nitrate and phosphate pollution, brackish and saline water depleting ground water siltation of water basins	Uncontrolled irrigation and fertilizer application, improper irrigation and water management, excess ground water withdrawals, soil erosions (run off)

(Contd.)

Resource	*Degradation/Depletion*	*Casual Agricultural and Production Activity*
Air	Increased concentration of methane, ammonia and nitrous oxide (Global warming) adversely affecting photo synthesis	Increased paddy cultivation, excessive animal population (unproductive stray animals). Increased use of organic fertilizers
Forest	Denudation of sand, poor crown cover, air pollution, and reduced photo synthesis	Shifting cultivation, deforestation, poor forest management, reduced forest area

Source: Impact on Health and Environment, 1993.

Land, water, air and forest are basic resources and they are often subjected to typological degradation/depletion over a period of time causing insecurity to human and causative factors for this process. These variables can be selected for studying a micro region and how macro changes are detectable, sometimes they could be done with human eyes. So, remote sensing methods/GIS/GPS helps to estimate the loss to planet earth. The geo-morphological study may be clubbed into RS/GIS with aerial photographic method for interpretation is much desired to estimate and mapping them over a period of time.

Air Pollution and Health Hazard

> "Recently emphasis has shifted on the effect of different types of pollution on human health instead of pollution of an area, city, or region. This issue is gaining more attention in the research field of Indian Geographers".

J.M. May, was the first medical geographer who mentioned the work in 1950 and the American Geographical Society published the Atlas of Diseases in three volumes.

(1) The Ecology of Diseases, 1958,
(2) Studies in Diseases Ecology, 1961, and
(3) Ecology of Malnutrition in the Far Near East, 1961.

Because of the significance of micro level study, the author decided to take up a detailed study of air pollution and its effect on human health.

Methodology

(1) Primary and Secondary data have to be collected from various sources,
(2) Secondary data have to be obtained from Municipal Corporation,
(3) State Pollution Control Board,
(4) Census Office,
(5) Primary data has to be generated through personal visit/ survey or sample study of the affected sensitive areas of Air Pollution, and
(6) The study is purposive in nature.

Climate and Agriculture

(1) Normal annual rainfall and analysis of temperature distribution over a period of time (two points of time),
(2) Maximum, Minimum and Mean annual temperature, and
(3) Seasonal distribution of rainfall.

Method of Analysis

(1) The Coefficient of variation in rainfall (calculation),
(2) Trend of Rainfall – to identify this pattern five year/10 year moving average and by the least square method, and
(3) The correlation coefficient measure shows positive (+) or negative (–) relationship between rainfall and net sown area.

(2) Climate Change Vulnerability of Agriculture

Research Methodology

Step 1: The secondary data sources include historical data over 30 years from past to present period (t_1 and t_2) of WMO standard in respect of rainfall, maximum temperature, minimum temperature and rainy days from Indian Institute of Tropical Meteorology (IITM), Pune.

Step 2: The temporal and spatial analysis of data has been done by using moving average (Five years) technique. This tool is used to find out the trends of temperature and rainfall variation.

Step 3: For construction of vulnerability index, information related to population density, rate of population growth, level of literacy, landholding size and area under irrigation have to be collected from Census.

Step 4: Information related to forest cover and water resources/ waterbodies have to be extracted from Survey of India toposheet.

Step 5: Data on the rate of food production have to be obtained from District Agricultural Statistics Handbook.

Step 6: Information about ground water position have to be taken from Central Ground Water Commission.

Step 7: Data related to poverty, per capita income and income source have to be taken from Government of India web site/if it is APs, Government of Andhra Pradesh.

Analysis

Five Year Moving average technique is to be used in order to find out the trends of temperature and rainfall variation over the time period $(t_2 - t_1)$

Step 8: The vulnerability index as developed by M.S. Swaminathan Research Foundation, Chennai on, 'vulnerability assessment and enhancing the 'adoptive capacity' project under the climate change programme. The index is developed as a statistical tool to assess the existing vulnerability of a region to climate change in terms of the correct level of degraded resources.

Result: The results from these tools and techniques have to be represented using various diagrams, mapped and using computer aided cartography and are to be interpreted for valid conclusion.

(3) Methodology

"Climate change scenarios are a plausible and often simplified representation of the future climate, based on an internally consistent set of climatological relationships. LARS –WGI is a stochastic weather generator based on the series approach. (Racsko, *et al.*, 1991) with a detailed description being given in Semenov *et al.*, (1998) LARS-W.G. produces synoptically time series of maximum and minimum temperature, precipitation and solar radiation. The weather generator uses available observed daily weather for a given site to determine or set of parameters for probability distributions of weather variables as well as correlations between them".[2]

Identification of areas considered in the assessment of environmental impact (EIA).[3]

Keeping in view of the changing physical, economic (agricultural), social and demographic aspects, for environmental impact assessment studies, the following categories are classified and considered for the preparation of GIS data-base.

(1) Demography, (2) Housing, (3) Public Health as Morbidity, (4) Topography, (5) Climate, (6) Vegetation, (7) Economic activities, and (8) Ambient environment and related aspects.

EIA is essentially a complex and multi-disciplinary work, very few organizations and agencies are equipped to carry on the EIA, especially in case of a large river projects such as irrigation, Hydro climate and multipurpose dams.

"Environmental Impact Assessment is a designed tool to identify and predict the impact of a project on the bio-geographical environment and onwards health and well-being, to interpret and communicate information about the impact, to analyse site and process alternatives and provide solutions to shift out or abate/mitigate the negative consequences on man and environment".

The basic requirements for any consultant in EIA are:

(1) Knowledge or awareness of relevant rules, regulations, legislation and other legal requirements including standards for environmental parameters. India is concerned, people should be familiar with the EIA notifications of the Ministry of Environment and Forests, GOI, Rules and regulations of GOI;
(2) Detailed Projects Report (DPR) which provides all the informations about the dimensions, scale and magnitude of the project based on the DPR, the consultant should judge whether the project needs prior environmental clearance (scope of the project) and a feasibility study;
(3) Environmental Impact Assessment (EIA) may be comprehensive or rapid, depending on the scope of the project as per the relevant rules;

(4) If a project has significant environmental impacts, a mitigation or management plan in the form of Environmental Management Plan (EMP);
(5) Public hearing or public consultation if required, during scooping, and pre-feasibility, the expert committee decides whether public hearing is required; and
(6) Submission of the final EIA including EMP and presentation of the square for environmental clearance. The consultant should be able to satisfactorily answer all the queries of experts and provide any further data or classifications or informations or technical details required (Prof. K.B. Reddy) E.mail: *Prof.KBR@gmail.com.*

Oceonography

Methodology

(1) "Sea level rise and coastal vulnerability—An Assessment of Andhra Pradesh coast. Thorough Remote Sensing and GIS" the entitled study by B. Hemamalini and K. Nageswara Rao have presented a study of coastal vulnerability index, was prepared based on five variables. They are: (1) Coastal geomorphology, (2) Coastal slope, (3) Beach Changes, (4) Mean spring tide range end, and (5) Significant wave height.

(2) Methodology

Crop yield production using remote sensing, GIS and field studies in a case study of a part of Nagarjunasagar Command Area, Andhra Pradesh (India) by Y. Rama Mohan.

Attempted by the author, by availing the high resolution satellite data, implementing GIS and collection of GPS points to assess the crop pattern and to involve a widely acceptable model to suggest and predict the crop yields, the indigenous high resolution fused data consisting of IRSIC/ID and IRSPC LISS-III satellite data are availed for two seasons of (Kharif and Rabi) to generate the necessary information.

The micro information enables to analyse for the evolution of crop yield production, the spectral behaviour of selected crops like paddy, Groundnut, tobacco and sugarcane, are critically analysed and utilized in the development of training sets, in the process of evolving the model.

Result: The output model compared with that of scientific observations, found that the outcome of the model is up to the satisfaction.

E.I.A, examines both beneficial and adverse consequences of the project design and implementation of the project.

Methodology

(1) To prepare drainage network,
(2) Present land-use/land cover,
(3) Hydro-geomorphology,
(4) Slope, and
(5) Soils.

Tool

Satellite Data and Other Conventional Methods.

Result: Integrating this all information, various drought proofing works were recommended for implementation.

(1) Effective implementation and people's participation in the watershed management programme, there is control on soil erosion, reduction in siltation of the tanks, rise in ground water levels, changing in cropping pattern and increase in the economic returns to the farmers of the Narsapuram village (Anakapally district). (J. Narendra Kumar and G. Gopal).

Eco-climatology is the study of relationship between natural vegetation and climatic parameters like rainfall, temperature, potential evapo-transpiration actual transpiration, water deficit and water surplus and data has been utilized to analyse monthly water balance elements like precipitation, potential evapo-transpiration, actual transpiration, water deficit and water surplus adopting *Thornthwaite and Mather* (1955) *Water Balance Method*—Water balance method and ecological degradation studies by M. Sambasiva Rao, K. Somanna and B. Anjaneyulu are important in this direction.

7.2 RESEARCH INPUTS AND OUTPUTS IN ECONOMIC GEOGRAPHY

Economic Geography is subdivided into two forms. They are

(1) Agricultural Geography, and
(2) Industrial Geography.

Data collection in relation to conceptual trends and methodologies are identified and selective variables in this major branch of Geography is presented in this part, i.e., Research inputs and outputs in Economic Geography.

Agricultural Geography or Geography of Agriculture is essentially a study of agriculture in relation to prevailing geographical conditions. It is stated that Indian Agriculture is a gamble with monsoon. It is understood that "agriculture" has its pertinent and permanent roots associated with monsoons. Land is a source and resource to generate wealth from this (land) wealth to a farmer. Farmer's conditions are well conditional by monsoon rainfall, temperature, aridity and humidity with varying pedagogical characteristics which suits to grow crops and it may be able to give maximum returns with the minimum cost factor.

Physical factors, biotic factors, a biotic factors, economic factors, social and cultural factors will have a combined and multiple effects on Indian agriculture.

1. Physical factors often play an important role. For example, the influence of rainfall on agricultural productivity.
2. Biotic factors have a telling effect on agriculture and crop patterns.
3. Economic factor play an important and key role on land utilization and cropping pattern.
4. Cropping intensity depends upon the choice of the farmer as well as crop return, depending on economy of scale, i.e., and related to cost benefit analysis in agriculture.
5. Cultural factors play a dominant role, predominantly among the Indian farmers.
6. Technological inputs are the 'central and key' factors for the origin and growth of crop culture.
7. Agricultural diffusion is apparent rampantly not only at the local or regional level but also at the State/centre/global level.

Agricultural land-use studies, cropping pattern, crop productivity, technological inputs, crop outputs, spatial variations in land-use, crop use and crop productivity are the phenomena which is also cited with the land as well as farmers agricultural cost/agricultural outputs are two important aspects of a farmer to decide the choice of crop on a particular 'land'. It is said that land cadre is the strength to the farmer and further sowing of a particular crop. For any researcher it is a challenging point to bear in mind, "Is there any significant change in land-use pattern and conversion of land from the categories of forests, barren and waste land to the agricultural land".

"Is there any noticeable increase in the net sown area? Is there any fast change in land-use due to the expansion and intensification process of agricultural land-use have been accelerated by intensifying the technological inputs."

"Are there any market forces and expansion of road network which facilitates intensification of agricultural practices".

Land-use changes could be attributed to demographic expansion, general land-use, intensification, technological inducement and land-use is subjected to spatial variation over a period of time.

Hypotheses

(1) Related to land-use and population nexus as increasing population growth and population pressure intensify the agricultural land-use practices (Boserup, 1965); and

(2) Agricultural practices are intensified and productivity is raised by using modern agricultural technology (Binswanger, 1968). "Technology oriented land-use changes is still valid for the changing pattern of agricultural land-use".

Case 1

Step 1

Land-use categories are prepared by Directorate of Economics and Statistics. Broadly, the following are the general land-use categories.

(1) Forest land; (2) Land put to non-agricultural use; (3) Barren and uncultivable land; (4) Pasture and other grazing land; (5) Land under miscellaneous, e.g., Trees, grooves, etc., (6) Cultivable waste land, (7) Fallow land other than current fallow, (8) Current fallow, and (9) Net-Sown Area (NSA), (10) Cropping Intensity (CI), and (11) Total Cropped Area, viz., per agricultural worker.

Step 2

The average change is to be calculated by taking difference of ($t_2 - t_1$), latest year minus previous year of investigation, dividing it by number of years.

(*Source*: Land Utilization Statistics, Directorate of Economics and Statistics)

Step 3

1. Land-use categories should have to be represented by double bar diagram (against Area percentage).
2. Land-use categories against average annual change percentage).

Step 4

Changes in crop intensity in response to technological factors. Such as (1) Crop intensity, (2) Area under HYV Crop (Paddy), (3) Area under irrigation potential utilized, (4) Area under irrigation potential created, and (5) Number of tractors.

Step 5: Technological Inducement

(1) Area under HYV (paddy),
(2) Irrigation potential utilized (ha),
(3) Irrigation potential created (ha),
(4) Fertilizer (kg), and
(5) Tractors (nos.).

Total area/per 100 ha. NSA/Average annual change (per 100 ha. of NSA)

Step 6: Socio-Economic Indicators by Sex over a Period of Time (T_2-T_1)

Case 2: The Objectives are such as "To Examine the Evolving Land-use Patterns

- To analyse changes, spatial and temporal land-use with a case in point.

- To assess the nature of development in the context of local economic development (micro level)

Step 1: Population growth (t_1, t_2, t_3, t_4 and t_5) – variations

Step 2: The maps have to be generated from FMB (Field Measurement Book) using data building technique.

Step 3: The generated data have to be geo-referenced, coded with and of data dictionary. The consistency of the digital data in terms of 'area' has been validated using S.D. technique

Step 4: Spatial data have to be linked to general land use.

Step 5: Generalized land-use maps have to be prepared for the time period t_1, t_2 and t_3 (each decade) and a detailed land-use map has to be prepared and the change detection analysis is to be performed. Change detection analysis has to be used for identifying the temporal variation such as ($t_2 - t_1$); ($t_3 - t_2$) and ($t_4 - t_3$)

Step 6: Softwares to be used for preparation of maps are "Auto CAD, Ave/into, Ave view GIS".

Case 3: "Land-Use Planning Model for Srikakulam and Environs: An Urban Dynamic Process in a Rural Setting" —27th International Geographical Union, Washington, 1992, IGU Congress.

"Methodology has been deduced from the first law of thermo-dynamics, an urban dynamic model is presented and the efficiency or competency of a person is important for one self engaged in particular profession and nature of work performed by the individual. Thus, the efficiency of an individual depends upon the individual himself but not the distance travelled by him to attend the work spot, just as the internal energy of a thermodynamic system depends upon its state only but not upon the path followed by it to reach that state. It is concluded that for an efficient person distance function is not a criteria for attending the duty other than of his own locality.

Methodology used: Gravity Model, Christaller's Central Place Theory and Heedays land-use model are discussed in detail" (Author K.L. Narasimha Murty) [*Ref.* Readings in Development Studies, p. 257].

Case 4: Production and Export (Item) in Indian Agriculture

Specific Objectives

(1) To study the trend and growth of area, production and yield of paddy/sugar cane/ginger, etc.
(2) To analyse the export status of item in India in respective value, forms and market of exports.
(3) To study the constraints in the export of ginger from India and cfforts for increasing the exports.

Methodology

Simple statistical tools like growth rates and percentages are to be used for analysis. The compound growth rates of area and production of ginger are to be worked out by fitting the exponential function given below.

$$Y = a + bt$$

Y = the dependent variable (area and production)
a = constant
b = $1+ r$, regression coefficient
v = $(b - 1) \times 100$, CGR in percentage

Step 1. Area, production and yield of crop
1. Area (hectares) 2. Output (MT) and yield (kg/hectare)
Step 2: Area (hectares) – percentage; output (MT) – percentage
Step 3: Trend in export (year/quantity (MT): value (Rs.)

Case.5: Regional Imbalances in Food Security

Specific Objectives

(1) To show the regional pattern of food availability, food stability and food security in terms of a large number of indicators.
(2) To assess the underlying reasons for regional imbalances in food security.
(3) To suggest remedial measures to solve the problem of food insecurity in the study area.

Methodology

Step 1: Food availability in terms of kg per head per annum has to be worked out by finding out the total production of various items relating to food and vegetables grown in each district and dividing it by total population of the district.

Step 2: Availability of food in calories has to be calculated by connecting the physical production into calorific value in each district per head per day taking into consideration to the age and sex composition of population in the study area and their specific requirements.

Step 3: Food stability and food accessibility have to be worked out in terms of a large number of indicators including percentage of population above the poverty line, the percentage of working population to the total population, storage capacity, length of surface road and urbanization, etc., using the technique of standard score additive model ('Z' score). The 'Z' score is a linear transformation of original data in such a way that its mean becomes '0' and standard deviation becomes unity, for observation 'I' on any variable. The standard score is given by:

$$Z_i = \frac{x_i - \overline{x}}{S}$$

Where x is the original value of an observation.
$\overline{x}$ is the mean for the variable.
S is the standard deviation.

The overall situation or composite index of food security may be expressed as

$$SE_i = \sum_{i=1}^{m_{10}} z_{ij}$$

Where SE_i = standard for food security.
m = is the number of variable that is 10 in the present study.
ij = is the magnitude of indicators for district j
Zij = is the standard score of variable (i) in the district j.

Agricultural Geography

H. Nagaraj (Mysore) here deals with the study of spatial variations in cropping patterns and crop-combination in relation to fluctuation of environmental factors. Its aim is to *find correlation* between crop combination and environmental differences and also to find out influence of environment, socio-economic factors in the distribution of crops and crop combination, crop ranking and crop changes. The regional character of crop combination and regional domestic crop has been determined on the basis of the relative strength of various crops and ranking them according to the percentage of cropped area occupied by each crop.

Weaver's Method

Food Security Standard Statistical Method like, Composite Z-Score, to find out Relationship between variables of food security and agriculture, Karl Pearson's technique of correlation coefficient and Students-T- test technique has been adopted.

Regional Imbalance

Food security has been resorted low in those areas where agriculture development is high. It is due to highest areas of agriculture development dominated by cash crops while food availability has been recorded low in these areas which is a backbone of food security.

In this chapter there is an attempt of determinants and consequences of outmigration and relationship between rural outmigration and various development variables by applying *logistic analysis* with dynamics of *spatial* economy.

Factor Analysis to measure Economic Development

The 'Z' score method has to be applied in order to assess the level of "Higher use of fertilizers and pesticides to maintain yield levels which do not help rather results in so called second generation problem of ground water contamination and environmental degradation".

Agricultural Geography

H. Nagaraj (Mysore) here deals with the study of spatial variations in cropping patterns and crop-combination in relation to fluctuation of environmental factors. Its aim is to *find correlation* between crop combination and environmental differences and also to find out influence of environment, socio-economic factors in the distribution of crops and crop combination, crop ranking and crop changes. The regional character of crop combination and regional domestic crop has been determined on the basis of the relative strength of various crops and ranking them according to the percentage of cropped area occupied by each crop.

Regional Disparities of Infra Development

"The absence of proper planning for infrastructure sources can be identified as one of the important factors causing regional disparities. The technique of GIS have provided endoscope for designing strategies for infra development.

The main objective of GIS → capturing, storing, checking, integrating manipulating, analyzing and displayed data which are spatially referenced to the earth" (GIS software ILWIS) (Identifying the Causes)

Case 6: Factor Analysis

An intensive attempt but an extensive scale to analyse the trends of inter-Mandal disparities, construct composite indices of economic development and identify the backward, developed and highly developed Mandals with respect to physical, demographic, infrastructural and overall development. Development is a process with complexities and have multidimensional structure depending upon resource base in general and infrastructural facilities in particular, and it is felt necessary to have a group of indicators to measure it.

Objective

1. To determine the economic development level and disparities in Mandals, cross-sectional data related to thirteen indicators has been used. The limitations of this study is that the information on all 13 variables do not relate to a single year;

2. To construct composite indices of economic development; and
3. To identify the backward, developed and highly developed Mandals with respect to physical index, demographic index and infra index and overall development.

Methodology

All the 34 Mandals have been included taking district as the unit. The variables considered in this study are as follows:

X_1 - Total rainfall
X_2 - Area under Principal Crops (APC)
X_3 - Total Cropped Area (TCA)
X_4 - Area of Principal Crops Irrigated (APCI)
X_5 - Rural Electric Service Connection for Agriculture (RESCA)
X_6 - Total Population (TP)
X_7 - Total Population SCs (TPSCs)
X_8 - Total Population STs (TPSTs)
X_9 - Educational Institutions (total number of institutions) [TNEI]
X_{10} - Postal Service (total number of Post Offices) [TNP]
X_{11} - Veterinary Services (total number of Veterinary Centres) [TNVC]
X_{12} - Banking Services (total number of branches) [TNBS]
X_{13} - Medical Services (total number of hospitals) [TNH]

Source of Data

The data was collected from secondary sources from MRO of 34 Mandals relating to the selected variables.

Study

In this study, in order to know the structural components of economy to construct *composite* indices of economic development, principal component method of factor analysis which has been used thus seeks to resolve a large number of measured variables in terms of relativity to few categories known as factors, each factor explains the maximum amount of variable percentage. The greater advantage of this method is that it gives a mathematically unique solution. The factors are so obtained are principal components (Kothari, 1990).

Results and Discussion

(see and consult the titled book)[4]

Case: Measurement of Agricultural Efficiency: A Comparative Study

Methodology

An attempt has been made to measure the agricultural efficiency and regional disparities and their comparisons. The agricultural efficiency may be measured with the help of transformed standard score taking into account indicators belonging to five groups like rainfall or irrigation, yield, land use, demography and agricultural innovations. Regional disparities are measured with the help of *Kendal's rank order score* method. Considering 31 indicators belonging to social aspects, economic aspects and demographic aspects.

Agricultural productivity is the actual performance of the unit land in its yields. Although, there are many methods to measure agricultural efficiency, they suffer from one or other weakness. In view of this, it appears appropriate to make use of weighted transformed student score method.[5]

Nineteen indicators of agricultural efficiency are selected after a careful study of their relative importance and are classified into five broad groups:

I. Rainfall and Irrigation Indicators

1. Mean Annual Rainfall (cm);
2. Percentage of irrigated area to the net sown area;

II. Yield Indicators

3. Per hectare yield of rice (in tonnes);
4. Per hectare yield of ragi (in tonnes);
5. Per hectare yield of jowar (in tonnes); and
6. Per hectare yield of other cereals (in tonnes).

III. Land-Use Indicators

7. Percentage of net sown area to the geographical area;

8. Percentage of area sown more than once to the net sown area;
9. Percentage of area under rice to the district total;
10. Percentage of area under ragi to the district total;
11. Percentage of area under jowar to the district total; and
12. Percentage of area under other cereals to the district total.

IV. Demographic Indicators

13. Percentage of agricultural workers to the main population of workers; and
14. Percentage size of land holding by the agricultural worker (in hectares).

V. Agricultural Innovations

15. Percentage of drought animals to the total livestock;
16. Percentage of irrigation pump sets to the total of the district;
17. Percentage of power consumption to the total of the district;
18. Percentage of tractors to the total of the district; and
19. Percentage of fertilizer consumption to the total of the district.

The index of agricultural efficiency for a taluka can be expressed in the following equation:

$$\text{IAE} = \frac{T_{21}(w_1) + T_{22}(w_2) + T_{23} = (w) + \ldots + T_2 n(wn)}{N}$$

Where IAE = Index of Agricultural Efficiency of a taluka.
T_2 = Transformed standard score of an indicator.
N = Number of indicators.

The values of 19 indicators not only differ in their basic units but also in their relative importance. Therefore, it is necessary to convert those values in each of the 19 into comparable unit values. This can be done by computation of standard scores since standard scores are equal units of measurements and also their size is the same from one to the other distribution, so that they become useful tools in the present study.

The formula for the standard score is as follows:

$$\text{Standard score} = \frac{X - \overline{X}}{SD} = \frac{X}{SD}$$

Where X = any row score or unit of measurement in a distribution.

$\overline{X}$ = mean value of the distribution.

SD = standard deviation of the distribution.

X = (X–X).

The standard scores are equal units of measurements and hence can be manipulated mathematically. It should be noted here, changing distribution of scores 'Z' does not change the slope of the original distribution of scores. Since 'Z' scores are expressed in decimals and in some cases they are negative, they are rather number save to handle in order to overcome this difficulty.

Linear transformation of 'Z' value is made through such transformations make the scale larger, the negative scores are eliminated and decimals are also done away with. Transformed scores are obtained from the following equation.

Transformed standard score = (imagined standard deviation + marginal mean)

The imagined standard deviation considered is '10' and the imagined mean is '50'. In equal form, this becomes transformed.

$$\text{Standard score } (T_2) = \frac{X - \overline{X}}{SD} \times 10 + 50$$

Since all the indicators are equal in importance, it is necessary to assign weight to them.

$$\text{Weight is equal to mean} = \frac{mean}{SD} = \frac{X}{SD}$$

The values thus obtained are taken into consideration for classifying the talukas with regard to the level of agricultural efficiency. While classifying the talukas as very high, high, medium, low and very low agricultural efficiency regions, *mean* and SD formula is used.

Change

The percentage change of agricultural efficiency over a period of time $(t_2 - t_1)$ is worked out and mapped with the help of *quartile* method.

Measuring Agricultural Output

Land resources may be identified by the amount of output in a region was attempted by geographers, agricultural economists, and planners, etc. The following method is more useful where sustainable agriculture is being adopted.

For assessing the agricultural output, the following four means could be adopted in various instances:

(1) The yield of main crops with their remaining coefficients in various areal units following Kendall (1939) and Stamp (1960);
(2) The concept of weighted share of production where ratio of average yield per unit area is considered (Ganguli, 1938) (Spare and Deshpande, 1964) and Bhatia, 1967);
(3) The relative crop equality which provide the equal share profile and importance of a crop in a region as a whole. The concept was first introduced by Buck (1937) and was later reshaped by Singh, Jasbir (1972) using caloric significance of crops;
(4) Lastly, the concept of corporate index, was worked out by Pal (1963), modified by Bhat of Learmonta (1968) and later followed by Hussain.

Cropping intensity index (CII), i.e., relative weight of the share of double cropped land is calculated to divide the same with its regional average, has been composed into a single index of agricultural output, which may be symbolically written as:

$$Te^{1} = (\text{I Sy}^{3}, \text{I we,}, \text{I ci})\ 10^{4}$$

Cropping Pattern

Crop concentration or diversification is the result of diversity in geo-climatic, and topographic, location, socio-economic conditions as well as irrigational development, technological development and institutional set-up for future crop planning and orientation of farming on the right lines, the traditional rural agrarian economy needs to be changed and suitable crop associations should be evolved and the entire agrarian structure has to be transformed. For such a progressive agricultural planning, a study of crop concentration, crop regions, crop combinations and crop diversification is a prerequisite methodology in Agricultural Geographic Research.

Step 1: Identification of crop concentrations on the basis of percentage of individual crops to the total cropped area;

Step 2: Delimitation of primary or macro-crop regions on the basis of ranking of the crops;

Step 3: Identification of the basic crop combinations on the basis of Doi's method; and

Step 4: Examining the extent of diversification of the cropping pattern in the district on the basis of the Gibbs-Martin Statistical Method.

Quantitative-statistical techniques are more exact, scientific and objective. The "Minimum Deviation" method developed Weaver (1954) was the first attempt to determine the crop combination regions in the mid-west of the USA. This method has failed to determine crop combinations and showed anomalies in certain cases.

Doi (1959) followed Weaver's method but avoided the time consuming calculations by excluding 'N' from the formula and by introducing pre-calculated critical values. By making use of the table of critical values of Doi, the crop combinations in any region can readily and easily be determined. Doi's method seems to be better than the other in sense that it is simple and involves much less calculations while at the same time it gives a realistic picture of crop combination. This method can be employed to any region and the results obtained are comparable.

Gibbs Martin (1962) developed an index to measure diversification of employment in industries. This method provides a useful and suitable alternative index for measuring the degree of diversification in the cropping pattern in an area Dhabai (1979) applied Gibbs-Martin's index to measure the indices of crop diversification in Rajasthan which gave satisfactory results. The formula developed for calculating the index is stated as under.

$$\text{Index of Diversification} = \frac{\varepsilon x^2}{(\varepsilon x)^2}$$

Where 'x' is the percentage of the total cropped area occupied by each crop or average under individual crop. If the total cultivated area in region is devoted wholly to one crop showing specialization, the index value will be zero; and if it is evenly distributed among many crops showing maximum diversification, the index value approaches one. This method has an advantage over the other methods and are (i) The statistics of the crop need not be reduced into percentages; and (ii) the indices indicate the magnitude of diversification in direct proportion.

Gibbs-Martin Method

This method can be employed to measure crop diversification. For selecting the major crops the procedure followed in this method is as follows.

(i) Divide 100 by '*N*' number of crops grown in an area. Then consider only those crops as important if they cover an area larger than 100. The result of the crops, which are grown in an area less than 100 are ignored.

M.V. Reddy's Method

In Chittoor district, 8 crops are considered for the purpose of calculation. Only those crops which are cultivated in more than 100th per cent of the total cropped land in the district are considered as major crops. The rest of the crops are ignored. Later the index of diversification is calculated by using the formula.

Finally, in order to examine the temporal changes in the crop concentration, crop regions, crop combinations and crop diversifications, the study is made at two points of time namely 1972-73 and 1982-83.

Method: Centrographic Analysis—Sugar Industry: A case study.

Data: The related information and statistics have to be collected from secondary sources like National Federation of Co-operative Sugar Factories Ltd., New Delhi, Annual Reports.

Technique: Personal observation techniques is to be adopted in order to check the ground realities.

Methodology

"For the analysis of location patterns of sugar industry the Centrographic technique has been employed to study the point pattern, to find out central tendency and dispersion of any phenomena.

The mean centre $(\overline{X})$ is a measure of central tendency within a spatial distribution and it is calculated in a similar way to the mean of numerical distribution and it is calculated in a similar way to the mean of numerical distribution. The calculation of standard distance appears to be the simplest measure of geographical dispersion associated with the centre. This index is the equivalent to standard deviation (SD) in a numerical distribution and measures the degree of dispersion of points about the mean centre (X/'o') (Bachin, 1993) where standard distance SD is calculated using the following formula.

$$SD = C\sqrt{\frac{\Sigma fd^2x}{n}\left|\left[\frac{\Sigma fdx}{n}\right]^2 + \frac{\Sigma fd^2y}{n}\right|\left[\frac{\Sigma fdy}{n}\right]^2}$$

Where SD = standard distance of radius of circle.
c = class interval.
f = frequency of factories.
d = distance of each factory from mean centre.
n = number of factories.

The standard distance is the value that allows the dispersion of different point patterns to be objectively".

7.3 POPULATION GEOGRAPHY IS AN INPUT TO GEOGRAPHICAL RESEARCH FOR PRESENT AND FUTURE GENERATIONS

A brief birds eye view: "The discipline of geography is itself one of the young disciplines in India. In early days of development of geography in India, urban geography and agricultural geography remained relatively more favoured branches of systematic geography, courtesy a few senior geographer's of the yester years. The initial framework of India's population geography was provided by doctoral work of Gosal (1956) supervised by Glenn Therewartha. Subsequent study by using similar methodology were carried out under Gosal's supervision at the Punjab University since early 1960's (Krishan, 1969, Chandha, 1970, Mehta, 1971).

The physics of population and the philosophy of population, India has been consistent with philosophy of geography itself. Mostly, the predominant mode of enquiry employed in geography is the spatial perspective, population geography in India has consistently maintained a spatial perspective and stemming the focus into spatial distribution, spatial relationships, spatial interactions while dealing with any demographic phenomena (Gosal, 1984). Thus, spatial approach consistently constituted the methodological basis of geographic study of population in India. Growth of population geography in India has, its own demands, in conceptual as well as methodological fields.

"Facts to theory" approach has been followed that would be stating half the truth. There are a few attempts to use 'theory to facts' approach at doctoral approach over empirical approach. The two aspects are so interlinked and independent.

The use of theoretical approach is important upon comprehensive training of population geographers in the use of quantitative techniques which seem to have been mastered by sociology, economics, psychology, etc. For example, Multivariate analysis, factor analysis, and principal component analysis are only a few techniques which need to be mastered by geographers, who have to analyse situation which are becoming increasingly complex in the modern world. Model briefing hypothesis testing and theory formulation are the real challenging fields for population geographers of the coming years". [Chandna, 2010 (p. 21)].

According to Chandna, "the quantification and use of sophisticated techniques shall not only bring precision in interpretation but also shall help in the development of theories. Theories and models developed elsewhere in the world have their own spatial, cultural, and political-historical context. The demographic scene in India provides a rich laboratory to population geographers for the development of models and theories that fit into spatial cultural-political and historical context of India.

Two methodological issues are crept in. They are (1) core, and (2) peripheral issues.

The first category in concerned with (i) The distribution and structure of population, mortality, fertility and migration; (ii) what causes them to be as these are; (iii) why these change through space and overtime; and (iv) how and why these affect other socio-cultural economic, political and environmental issues. (2) At the periphery the issues that are central to other specialization such as urbanizations, social segregation, etc. Population geographers have restricted their studies to those attributes, the most important are the growth of population, literacy, sex composition, urbanization and working force.

Population sheets published by the National Atlas Organization using 1951 data apart, there have been only two such studies (Chatterjee, 1962 and Chandna *et al.*, 1985) for India as a whole. Chatterjee used 1951 Census data, Chandna and Kant used 1981 district level data. These two studies are spread over a period of 30 years on such a crucial attribute as distribution and density speak of the serious gaps in the seminal content of population geography in India.

Studies on the deprived sections of Indian society such as Scheduled Castes and Scheduled Tribes have also caught geographers imagination (Chandna, 1985; Chandna, 1989). Atlases of Scheduled Castes (Chandna, 1985) and of Scheduled Tribes (Raza, 1990) are welcome additions to Indian Population Geography.

More importantly referenced one are the volumes on demographic dynamism of India, on migration patterns in Punjab's region, on unemployment and underemployment in Punjab and on urbanization in India are welcome additions to the growing number of books in this field of specialization (Smita, 1995; Mehta, 1990; Krishan, 1985 and Bala, 1986).

It is observed that, "deficiencies of data continue to handicap geographers as they continue to shy away from field work, which is none too easy to conduct in Indian situations. Micro level studies are now picking up on a less speed and vigour and so are the theoretical formulations by Indian population geographers, challenges are serious, warranting equally serious efforts."

The author's primary aim is to train primary the students and scholars of geography as well as social science discipline, the spectacular changes are mentioned keeping in view of the literature and national level population deals with population facts and figures that have been prepared by Census of India organization, New Delhi. The changes in the variables that are used often over period of time and the same could be much appreciated by students and scholars belonging to population studies. In the following paragraphs, the details are stated as it warrants to be the author to present them in order to understand the changing patterns of collection of data, design and departure.

The 2001 Census considerably expanded the classification compared to the earlier censuses to include the following:

A. **Primary Sector:** Cultivator, agricultural labourer, hunters, and foresters and those engaged in animal husbandry, plantation, mining and quarrying.

B. Secondary Sector: Manufacturing, processing, servicing and repairs, whether carried out in the form of household industry or at a higher land and construction activity.

C. Tertiary Sector: Trade, commerce, transport, storage, communication and services.

The 'NSS employment and unemployment survey divides the entire population into three mutually exclusive and exhaustive sets, depending on either 'major time' or 'priority' criteria. These are:

(a) those working or being engaged in an economic activity;
(b) those not being engaged in economic activity but either making tangible efforts to seek work or being available for work if work is available; and
(c) those not being engaged in any economic activity and also not available for work.

2. To determine how much of an economy's population is contributing to the production of goods and services. Work Participation Rate [WPR] has been defined as the population of total workers in the population, expressed as a percentage.

$$\frac{\text{Total number of workers of all ages} \times 100}{\text{Total Population}}$$

			Total			Rural			Urban		
			Per.	M	F	Per.	M	F	Per.	M	F
5-9	30-34	80+									
10-14	35-39	ANS									
15-19	40-49	15-59									
20-24	50-59	60+									
25-29	60-69										
	70-79										

(i) A more accurate denomination may be the work participation rate for the population aged between 15-59 years or even the population above the age of 7+.

(ii) Age specific work participation rate by sex and residence, India: 2001:

Work Participation Rate in States/Union Territories.

Work Participation Rate by Sex and Residence.

Work Participation Rate by Religion.

Work Participation Rate by Literary.

Work Participation Rate by Scheduled Caste/Scheduled Tribe.

Comparison of Work Participation Rate in Census 2001 with NSSO 55th Round [July 1999—June 2000]

Determinants	Census Workers	NSSO Employed [U.S. (P.S+SS)]
Minimum days of work	1 day	30 days
Minimum Time	Not specified in 2001	1 hour per day *Note*: Some of the women and children would be out of labour force, when time criterion is applied
Unpaid Apprentices	Worker	Out of labour force (student)
Temporary Sickness	Included in the count for days	Not included in count for days
Gathering of forest Produce for Consumption	Some workers such as those collecting minor forest produce while others such as those collecting fuel wood, fodder, grass, etc., for home consumption are non-workers	All are taken as employed
Persons who did not work any time but was seeking/ available for work	Not a worker	Excluded from the employed but included in labour force as unemployed (constituted nearly 1 per cent of population in 1999-2000)

Employment Rates by US (PS) states as per NSSO 55th Round and percentage of main workers as per 2001 Census.

	NSS-1999-2000				Census 2001			
US (PS) Status	Industry	Persons	Male	Female	Location	Persons	Male	Female
11-51	01-99	36.5	52.0	20.3	Rural + Urban	30.4	45.1	14.7
11-51	01-99	38.0	52.2	23.3	Rural	30.9	44.3	16.6
11-51	01-99	32.4	51.3	11.7	Urban	29.3	47.2	9.4

Note: Status codes 11 to 51 cover different types of work.

Source: Appendix Table.1, Page. A1 to A5, NSS Report No. 458, Part-1 and Table B-1, Census of India, 2001.

Labour force Participation Rates by US (PS+SS) status of 55th Round NSS and Similar Rates from 2001 Census.

US (PS) Status	NSS 1999-2000				Census 2001			
	Industry	Persons	Male	Female	Location	Persons	Male	Female
11-81	01-99	40.6	54.0	263	R + U	43.5	55.7	30.4
11-81	01-99	42.3	54.0	302	Rural	45.2	55.2	34.5
11-81	01-99	35.4	54.2	147	Urban	39.2	56.8	19.6

Note: Status codes 11 to 81 cover both the employed and those who were not employed but were seeking available for work.

Source: Appendix Table-2, Page A-10 to A-18, NSS Report No. 458, Part 1 and Table B-1, Census of India, 2001.

[i] Total population and main workers in India, 2001.
[ii] Total workers/main workers/total persons/male/female.
[iii] Total/Rural/Urban.
(1) Main workers in the country.
(2) Main workers in States/Union Territories.
(3) Main workers in rural areas.
(4) Main workers in urban areas.
(5) Main workers by sex in the States/Union Territories.
(6) Marginal workers in the country.
(7) Differential among marginal workers by sex.

Analysis of the Industrial and Occupational Profile of the Workers

(1) Composition of Workers

Distribution of total workers by sex and residence, India, 2001:

(i) Total workers ....
(ii) Cultivators ....
(iii) Agricultural labourers ....
(iv) Household industry workers ......
(v) Other workers

Number of Main Workers, 1999-2001

Category

(i) Main workers
Agricultural sector
Non-Agricultural sector
Marginal workers
Agricultural sector
Non-Agricultural sector
Total workers
Agricultural sector
Non-Agricultural sector:

1. Area and population:
   1.1. Area and population by States (Census, 1991).
   1.2. Decennial growth of population by census.
   1.3. Estimated birth rates, death rates and Infant mortality rates by residence.
   1.4. Expectation of life.
   1.5. Estimated mid year population.

Slum Population India – Census of India – 2001

Distribution of Population Size of Cities/Town & Share of Slum Population

Pop.Size of cities/ Towns [in Lakh]	*No.of Cities/ towns*	*Percentage of cities/ Towns [in Lakh]*	*Per- centage*	*Slum Population In cities/ Towns [in Lakh]*	*Per- centage*
40+	5	350.8	19.0	110.6	26.00
20-40	8	212.4	11.5	37.6	8.80
10-20	14	170.3	9.2	28.8	6.80
5-10	42	310.5	16.9	58.1	13.70
1-5	309	605.1	32.8	139.4	32.70
Less than 1	262	194.4	10.6	51.3	12.00
Total	640	1843.5	100.00	425.8	100.00

Population Profile

	Population
India	1,028,610,328
Rural	742,490,639
Urban	286,119,689
State/UTs Reporting Slum	283,741,818
Towns reporting slums	183,352,421
Slum	42,578,150
Non Slum (640 cities)	141,774,271
Slum – 42.6 million (15%)	
Non-slum – 243.5 million (85%)	

Source: Slum population, Series-1, India, Census of India, 2001.

$$\text{Index of Agricultural Production} = \frac{\Sigma\left(a_{ij}\, y_{ij}\, p_{io}\right)}{\Sigma\left(a_{io}\, y_{io}\, p_{io}\right)} \times 100$$

$$\text{or} = \frac{\Sigma\left(p_{ij}\, p_{io}\right)}{\Sigma\left(p_{io}\, p_{io}\right)} \times 100$$

Where a_{ij} = area under the i-th crop in the j-th year

a_{io} = area under the i-th crop in the base year

y_{ij} = yield percentage of the i-th crop in the j-th year

y_{io} = yield per hectare of the i-th crop in the base year

p_{io} = price per unit of the i-th crop in the base period

p_{ij} = production of the i-th crop in the j-th year

p_{io} = production of the i-th crop in the base period

Reference: (i) Statistical Abstract, India, 2001, Agriculture.

(ii) Pattern of land utilization ('000 hectares)

Classification of reported area

1	2	3	4	5	6	7
Year	Population for land utilization statistics	Forests	Not available for cultiva-tion	Permanent pastures of other grazing land	Land under Crops and Grooves (not inclu-ded in Net Area Sown)	Waste land

Fallow land – Classification of reported area covered ('000) hectares

	8	9	10	11	12
Year	Fallow land other than current fallows	Current fallows	Net area Sown	Area sown more than once	Total cropped area

Statement – 4

Total population — 'All religious communities' 2001 urban/rural. Classification/Hindus/Muslims/Christians/Sikhs/Buddhists/Jains/ Others – Total population

Population of Towns (by classes) and villages (Census-2001)

State				
No.	173			Class
	Total Population		100,000 and above	1
			50,000-99,999	2
			20,000-49,999	3
			10,000-19,999	4

(Contd.)

State				
No.	173			Class
			5,000-9,999	5
			Under 5,000	6
Inhabitated				
Uninhabitated				
Census of India, 2001		-	Marital status and Age and Marriage	
		-	An Analysis of 2001 census data	
Census of India – 2001		-	District level	
			Estimates of child mortality in India. Based on the 2001 census data	
Census of India – 2001		-	A to Z of housing census Summary Results on Houses, Household Amenities and Assets – India Data Dissemination wing O/R/, India 2A, Mansingh Road, New Delhi	

The primary Census abstract for slum and total urban areas – 2001

Andhra Pradesh

Item 1	Sex pop. Male Female	Slum	Urban
1. Total Household	P.M.F		
2. Total Population (in it int. & Houseless Population)	P.M.F.		
3. Population in the Age group 0-6	P.M.F.		
4. S.C. Population	P.M.F.		
5. S.T. Population	P.M.F.		
6. Literates	P.M.F.		
7. Total workers	P.M.F.		

(*Contd.*)

8. Main workers	P.M.F.
(i) Cultivators	P.M.F.
(ii) Agri-labourers	P.M.F.
(iii) Household Industry	P.M.F.
(iv) Other workers	P.M.F.
9. Marginal workers	P.M.F.
(i) Cultivators	P.M.F.
(ii) Agri-labourers	P.M.F.
(iii) Household Industry	P.M.F.
(iv) Other workers	P.M.F.
10. Non-workers	P.M.F.

Sex Ratio: Sex ratio has been defined as the number of females per 1000 males in the population. It is expressed as number of females per 1000 males.

(1) Sex Ratio $= \dfrac{\text{Number of females}}{\text{Number of males}} \times 100$

(2) Child Sex Ratio (0—6 years)

$$= \frac{\text{Number of female children } (0-6 \text{ years})}{\text{Number of male children } (0-6 \text{ years})} \times 1000$$

(3) Literacy Rate $= \dfrac{\text{Number of literates}}{\text{Population age 7+}} \times 1000$

(4) Work Participation Rate (WPR)

$$= \frac{\text{Total workers (Main \& Marginal)}}{\text{Total Population}} \times 1000$$

Population Change and Urban Land-Use

Relationship using an integrated remote sensing and GIS approach for the study of population change and urban land use.

> "In developing countries, rapid population growth has meant a decline in the arable land per capita and switch over to industrial, residential, and commercial land-uses. If current trends in population continue, it is estimated that by 2050, the amount of arable land will be just over 1/10th hectare per person".

"The pressure of continuously growing metropolitan city also changing the town and its surrounding neighbourhood. The objective should be to investigate the changes in land-use pattern which have been occurring in the past for few decades and has time to associate them with population growth is visible in urbanization and industrialization of the countryside. For example Greater Hyderabad Metropolitan Corporation area (GHMC); CMC, Chennai; KMC (Kolkata Metropolitan Corporation) could be taken up by using latest methods in geographic research.

Methodology

While undertaking such studies, *multi-temporal RS and GIS* data sets have to be used. Urban land-use mapping have to be carried out using multi-temporal data. The spatial and temporal changes in various land-uses and it may be observed that, the surrounding region will grow at a faster rate over a period of time. For example, 1991-2001, 2001-2010 and so on. It helps to identify a complete metamorphosis in land-use.

Population Geography

Geographical Analysis Population Pressure — An Approach to Identification

1. Population pressure on land by considering a few but important indices like growth rates of population and food grains, cultivated land and foodgrains availability per head to visualize the magnitude of the problem.
2. "Over population could be observed to exist where there is too large a population in a given area for the actual or potential resources to support (Hemmond, 1985, p.32). The symptoms of such a situation are low income group, high level of unemployment, declining living standards, shrinking of land under cultivation and much less availability of food than what it is required, famine/drought, etc.

Method 1

Population pressure could be measured by applying the following formula:

$$I = \frac{P_1 - P_2}{A}$$

Where 'I' is the index of pressure;
'P_1' is the population size capable of being supported by the resources in a given region or in a country;
'P_2' is the actual population size in that region; and
'A' is the total area of the country.

Method.2*

$$I = \frac{P_1 - P_2}{F}$$

Where 'I' is the index of pressure;
'P_1' is the ideal population size;
'P_2' is the actual population size; and
'F' is the quantity of food available in a region.

Geometric Progression (Mandal, 2000, p. 316)

$$Y^{arn} - 1$$

Where Y is the year selected for progression,
a is the first year,
r is the rate of population growth,
n is the interval of the year, and
-1 is the constant.

* Micro data for Andhra Pradesh Population could be analysed through a standard statistical methods such as:

Curve Fitting

Let (x_2, y_2); $z = 1, 2, n$ be a given set of 'n' pairs of values, x being

* This analysis has been done by Dr. C.V.S.S.R.L. Hanumantha Rao, Lecturer in Statistics. M.R.A. College, *Viz.* Vizianagram, Andhra Pradesh is associated with the analysis of this Case Study on Micro Data for Andhra Pradesh Population.

independent. Variable and y the dependent variable. Fitting of curves to a set of numerical data is of considerable importance theoretical as well as practical. In practical statistics it enables us to represent the relationship between two variables by simple algebraic expressions. It may be used to estimate the values of one variable which correspond to the specified values of the other variables.

Best Fit

To a set of n points (x_2, y_2); $z = 1, 2, \ldots\ldots$ n given equation represents a family of curves for different values of the arbitrary constants of the equation. The problem is to determine arbitrary constants so that the curve of the best fit.

Legender's Principle of Least Squares

Minimizing the sum of squares of the deviations of the actual values of '*Y*' from its estimated values (*Y*) i.e.; $\sum_{i=1}^{n}\left(Y_1 - \hat{Y}\right)^2$ is

Minimum

The difference between the observed values page (2) known as residual or error of estimate and the task is to minimize these results. Since these differences may be positive or negatives, it is more convenient to make the sum of squares as these residuals minimum.

Fitting of Straight Line

Let us consider the fitting of a straight line

$$Y = a + b\,x \qquad \ldots [1]$$

To a set of *n* points $\left(x_2, y_2\right) \ni 2 = 1, 2, \ldots n$

Equation (1) represents a family of straight line for different values of the arbitrary constants '*a*' and '*b*'. The problem is to determine '*a*' and '*b*' so that the line (i) is the line of best fit.

According to the principle of least squares one have to determine '*a*' and '*b*' so that

$$S = \sum_{i=1}^{n} (y_1 - a - bx)^2 \text{ is minimum}$$

From the principle of maxima and minima the partial derivatives of S with respect to 'a' and 'b' should vanish separately, i.e., differentiating 'S' with respect to 'a' partially and equating to zero, we get

$$\frac{\partial s}{\partial a} = 0 = -2\sum_{i=1}^{n}\left(y_{\text{i-a-bx i}}\right)^2$$

Differentiating 'S' with respect to 'b' partially and equating to zero, we get

Equations (2) and (3) are called the normal equations for estimating 'a' and 'b'.

All the quantities ruined in equations (2) and (3) to estimate 'a' and 'b' can be obtained from the given set of points (x, y); $2 = 1, 2, \ldots n$, with the values so obtained equation (1) is called straight line of the best fit to the given set of points, i.e., $Y = \hat{a} + bx$: where a and b are the estimated values of 'a' and 'b'

Q. Fit a straight line to the following data x as independent variable. The population of Andhra Pradesh is given below in various Census Survey. Using this data to estimate the population of Andhra Pradesh in 2021. Population is in crores.

X	1971	1981	1991	2001	2011
Y	4.35	5.35	6.65	7.62	8.46

Year	*Population Y*	$X = x_1 - \frac{A}{10}$	X^2	Xy
1971	4.35	–2	4	–17.20
1981	5.35	–1	1	–5.35
1991	6.65	0	0	0
2001	7.62	1	1	7.62
2011	8.46	2	4	33.84
	$\varepsilon y = 32.43$	$\varepsilon x = 0$	$\varepsilon x^2 = 10$	$\varepsilon xy = 18.91$

we have the normal equations. Take $A = 1991$

$$na + b$$

Substitutions the values of εy, εx, εx^2, εxy in (1) and (2) we get

$$sa + bxo = 32.43$$
$$sa = 32.43$$

$$\therefore \quad a = \frac{32.43}{5} = 6.48$$

$$\therefore \quad \hat{a} = 6.48, b = 1.891 \qquad \therefore\ b = 1.891$$

$\therefore \quad \hat{a} = 6.48,\ b = 1.891$ substantially in $y = a + bx$

$$\therefore \quad \text{y} = 6.48 + 1.891 \ \frac{(x - 1991)}{10}$$

The population of A.P. in 2021 will be

$$Y = 6.48 + 8.91 \ \frac{(2021 - 1991)}{10}$$

$$Y = 6.48 + 1.891 \times \frac{30}{10}$$

$$Y = 6.48 + 1.891 \times 3$$
$$Y = 6.480 + 5.673$$
$$Y = 12.153 \text{ crores}$$

The expected population in 2021 will be 12.153 crores.

7.4 Geography of Settlements/Settlement Geography

'Geography of Settlements' is a prescribed paper at the post graduate course in geography in many Indian Universities. The title itself indicates that geographical environment has its own impact on settlement patterns and systems besides relating to the origin, growth and development of the settlements whether it is rural/urban or urban settlements. The terms and topics already covered the M.A. (Geography) Syllabus.

The geographic discipline, its place and significance, along with details on approaches, growth of studies and classifications with a detailed analysis relating to culture-historical perspective of settlements in the world is having a special emphasis on cultural realms, diffusion models, static temporal models and place name studies' are dealt. Village settlements, provides crystal clear view of the villages in general and Indian villages in particular, the meaning of site, situation and location/position, the village anatomy model, the internal social changes being experienced as well as study of sample villages.

Settlement geography deals with the types and patterns of settlements, giving their proper meaning, classification and causes, supported by study of sample villages and illustrations. The distributional pattern of settlements both qualitative as well as quantitative – such as spacing and dispersion and regional comparison, characteristics of new/pioneer settlements have been identified. Morphology is identified as comprehensive treatment warrants in depth explanation with context examples.

'Shape Analysis' also provides qualitative/quantitative approaches which facilitates to identify spatial distributional pattern of settlements. The case of settlements dewellings exhibits the theoretical and experienced based insight into the whole spectrum of various aspects of shape analysis or otherwise it is called as geometry of settlement pattern.

Growth centre concept and central place theory based identification of hierarchical, functional gaps as prerequisites for development planning but also of rural periodic markets.

In the twentyfirst century, integrated planning and rationalization of settlement having detailed account of social requisites, ecological basis and strategy of planning and feasible models of rationalization through spatial integration, functional co-ordination, efficiency of land-use and marketing as well as infrastructural development. Grass-roots and human settlement approaches would provide new insight to the disciplines related to the developmental planning."[7]

Urban and Settlement Geography

Urban places are considered as important focal points to generate growth in the field of economy. These centres vary from one to another over space, socio-economic, cultural and even political reasons have its impact on human life. Urban system is considered as a system, which could be identified by its growth and interaction of nearby settlements with the main settlement/settlement system over place/ spatial and urban settlement geographical studies channelled into seven sub-systems. They are:

(1) Evolution of urban systems, (2) Functional classification, (3) Critical study on urban fringe, (4) Urban ecology, (5) Demography and occupational structure of urban settlements, (6) Urbanization studies and methodological dimensions including applied urban geography.

Methodology for urban and rural settlements were developed by many geographers, economists, sociologists and demographers are quite important. Important methods are described in this branch of geography in the following paragraphs for its utility and application for future generation by researchers.

Urban Studies and Settlements

The logic of central place theory (Bunge, 1962) and the high correlation values between the size and functional range of settlements in empirical studies (Stratford, 1993) has helped in defining the system of settlement patterns and it is quiet important for carrying out research studies in this branch concept of population threshold. Generally, the minimum population that is necessary for the establishment of a particular service is known as the threshold size of that service. There is a general increase in the number of functions with the increase in size of a settlement. The exact form of the relationship between size of settlement and the number of establishment of each function differs from region to region (Tyne, 1971). A study of threshold size of service function mode by Harry and Gamison reveals that some services require a larger population for their support than the others and for the first time it was known, rather more precisely just how many people were, an average, require before a particular service could be provided in a settlement (Beny, 1958). Heggett and Gunewardana have tried to find a more redefined method using a modified (*Read Muench Technique* to calculate the median population threshold for settlement function (Heggett, 1968). This is to be known to strengthen the system model, which can also be applied for more precise decisions to implement by the governance.

Identification of the Centrality based Hierarchy

The centrality of a place in 'space' is defined by the amount of functions (*f*) needed to serve the urban region or urban space. "Centrality could be expressed qualitatively such as low, and high centrality, as well as quantitatively by centrality values which are secured by converting the functional base of a place into score on the basis of frequency and importance of the function centrality, however, depends upon the intensity of the central functions. The central functions follow a micro-schematic pattern. It is understood and deducible to arrive at a conclusion that the centrality of a place is defined by the number of functions that are needed to serve the population" (S.S. Hangeragi, 2005).

Methodology

(1) Measurement of Centrality by Godlund's Method

In the absence of any functional data, Godlund has attempted a classification of Sweden settlements on the basis of the capacity for track and service in the urban settlements. In this case he used *the relationship between the total population in a settlement and the number of persons employed in the retail trade and services.*

$$C = \frac{\overline{lc}}{p} \times 100 \qquad \text{... (i)}$$

where c is the regional mean index of centrality.

$\overline{lc}$ is the number of persons employed in retail trade and services.

P is the total population of the study region.

In this way, the centrality of every urban centre, all the urban centres whose index exceeds the regional mean are supposed to have a service area, higher indices being naturally associated with important urban centres.

(ii) Measurement of Centrality as per Davies Method

In this method the score for every function is assigned on the basis of the frequency of distribution of particular function. This method is analogous to the one used by Davies for South Wales. A score for a single unit of any function '*t*' was determined by the following equation.

$$C = \frac{t}{T} \times 100 \qquad \dots (2)$$

Where *c* is the score of function *t*.
t is the one unit of function *t*.
T is associate the total number of functional units of function '*t*' in the region.

The above formula is used here in the modified form by V.N.P. Sinha

$$C = \frac{t}{P} \times 100 \qquad \dots (3)$$

Where *c* is the desired centrality of the place.
T denotes the number of persons dependent upon tertiary service of a centre.
P is the total regional tertiary population.

Step 4: Mapping the relationship between centrality ranks and population ranks of urban centres.

Urban/Settlement Geography

Land transformation in relation to distance in developing economy

> "Urbanization process took off in developing countries in a highway but it is significantly different from the developed countries because of inherent social and economic differences. There will be spread concern in these countries regarding the expansion and land reorganization in and around urban centres."

Methodology

Step (1) The present study is based on both primary and secondary data.

Step (2) Urban land-use is to be prepared using remotely sensed data (Satellite imageries).

Step (3) These data along with extensive field checks and surveys provide accurate and up-to-date data.

Step (4) This helps in many ways, because urban areas are very dynamic and complex in nature and conventional data source were unable to cope with these data, as the informations were too much and they too get changed in a very short period of time. On the other hand their collection took a very long period.

Step (5) Mapping.

Study period (1974-2004), The urban sprawl mapping was done with the help of city guide map prepared by Survey of India (1974) on 1 : 20,000 scale and IRSID geo coded Panchromatic satellite analogue imaginary on 1:12500 scale acquired in 2001. The minimum delineation unit for both maps were 0.5 cm × o.5 cm. Further, the data were processed in GIS environment as GIS along with remotely sensed data provide the means for the rapid production of accurate and up-to-date data for planning.

Rural Settlements

'Settlements' have their structural features according to their size pattern and distribution besides urban settlements. Rural settlement study ranges from 1 to 7 categories. They are (1) Evolution studies, (ii) Environment and habitat, (iii) Spatial distribution of rural settlements, (iv) spacing and size of settlements, (v) hierarchy of service centres, (vi) village patterns (design and layouts), and (vii) morphology of rural settlements. Spatial variations in the distribution of rural settlements are important and this phenomena is with evaluating and mentioning as contributions over a period of time physical, economic, social and cultural factors will often play an important role in the regional distribution of rural settlements.

R.K. Rai, A.K. Singh, Jaipal Singh, Pandey, Jagat Narayanan, Ajay Kumar, H.P. Nayak, R.Y. Singh, Mahabir Chand Thakur and Jeelavi and M.A.K's studies are considered as the important contributions to rural settlement geography.

Spatial distribution of rural environmental structure (Bharat Saeed Khan and *et al.*, pattern analysis of rural settlement distribution (Nayak, D.R, 1985), spatial analysis of rural settlement and their types (Singh, K.N. and Minati Singh, 1980), Nayak, D.R.'s Study on spacing of rural settlements (1982), Guntia, R.'s study on comparative study of spacing of rural settlements (1985) are important and noteworthy.

Verma, R.P. and Pal, D.S. (1981) study on 'Hierarchy of Service Centres in the Kanpur, Of the Kanpur Metropolis, Bore Gowda, Penal Jaya Ramaiah P. (1981)'s study on 'Pattern of Rural Settlements in Karnataka, Zusti Bupinder (1981) 'Settlement Hierarchy in the Kashmir Valley', Prakash C.Tiwari and Dinesh C. Pandey's study on 'central place system and spatial organization, Soni M.L. (1981) study on, 'Hierarchy of Service Centres in the Lucknow, Nisarkhan's (1983) study on 'Hierarchy of Service Centres in India, Sayeed Ahmed Khan's (1983) study gives a new trend in settlement geography by identifying Hierarchy of Service Centres A New Approach, are important contributions in this branch of geography.

Chand, R. and Bhattacharya, G. (1981)'s contributions have special feature, "Determining typology of rural settlements in Hills, and for determining and measuring village shape are important (1985), Debala Devi, C. has her contribution, "Rural Central Places (1985) and assessment of von Thuhuen's land-use analysis by Jawaharlal Raina (1985) are most important contribution to identify the land-use pattern.

Finally, Mani, N.D. (1981) for identifying rural urban migration and its impact on urban growth centres and functional morphology of weekly markets relating to Bidar district, Karnataka by Yavagal, F. V. (1985) are quite important and solid contributions to this growing sub-field of geography and the recent contributions by applying GIS and remote sensing techniques are important.

Methodology – Settlement Geography

"Central Place Theory" (CPT) has been advocated by W. Christaller has assumed paramount importance not only in general but also globally. The essence of central place theory is that `central places work not only for its centres but also to their periphery areas. All consumers should necessarily have to avail of the services of the centre to reach that central place, people have to move to some more distance from his settlement for the upward movement (from lower to central place) in terms of economic terminology one has to spend his time as well as travel cost. The incumbent may not be able to reach that particular central place, keeping in view of the time constraint and cost factor. Even to provide the service to another settlement in hierarchy also, there is a lower limit (range of good) minimum population (threshold of population) required to locate such function in a particular context of settlement. Administrative principle, marketing principle and range of good are central and essential factors of Christaller's central place theory.

Methodology for determining Hierarchy of Rural Settlements

One way of adopting a strategy for determining the hierarchy of rural settlements in the approach, "Filling the gap strategy" and it is with this view, the study devotes to indentify rural service centre by classifying the settlements into hierarchical order.

Methodology and Data Base

Step (1): To determine the hierarchy of any settlement it is quite natural and essential to apply the concept of centrality of places

Step (2): The centrality of rural settlement as a service centre is indicated by the number of services it provides and proportion of these services supported by the population (Threshold of population).

The mathematical expression to calculate the centrality index is as follows:

$$C_1 = \frac{Os \times 100}{MX}$$

Where, C_1 is centrality index of the settlements.
Os is observed or weighted composite score.
Mx is maximum weightage scores.

Step (3): Rural Settlement Weightage scheme for parameters

1. Administration	Weightage
Adopted village panchayat	0.1
Village panchayat	0.2
Outpost	0.3
Police stations	0.4
Mandal Revenue Office	0.5
Municipal Corporation	0.6
Zilla Panchayat	0.7
Secretariat	1.0

2. Basic Infrastructure	Weightage
Well	0.1
Tap	0.2
Tank	0.3
Tubewell	0.4
Electricity Domestic	0.5
Community Hall	0.6
Children's Park	0.7
Theatre (Thetched)	0.8
Electricity Industrial	0.9
Star Hotels	1.0

3. Education	Weightage
Pre-Primary	0.1
Primary	0.2
School	0.3
Composite Junior College	0.4
Junior College (Boys)	0.5
Junior College (Women)	0.6
Industrial Training Institute	0.7
Teacher Training Institute	0.8
Degree College (Men/Women)	0.9
P.G. Degree College (Composite)	1.0

4. Transport	Weightage
Cart Track	0.1
Unmetallic Road	0.2
Metallic Road	0.3
Navigable water	0.4
Mandal Road	0.5
District Road	0.6
State Highway	0.7
National Highway	0.8
Railway Junction	0.9
Airport	1.0

5. Market	Weightage
Jatra	0.1
Urus	0.2
Church Feast	0.3
Seasonal Market	0.4
Weekly Market	0.5
Bi-weekly Market	0.6
Tri-weekly Market	0.7
Daily Market	0.8
Super Market	0.9
Regulated Market	1.0

6. Industries	Weightage
Household Industry	0.1
Floor Mill	0.2
Food Processing Unit (Biscuits)	0.3
Handicraft	0.4
Liquor Processing Unit	0.5
Oil Crushing Unit	0.6
Repair and Service Station	0.7
Small Scale Industry	0.8
Medium Scale Industry	0.9
Large Scale Industry	1.0

7. Medical	Weightage
Unregistered medical practitioner	0.1
Registered private practitioner	0.2
Public Health Sub-centre	0.3
Public Health Centre	0.4
Dispensary	0.5
Hospital	0.6
Mandal Level Hospital	0.7
District Hospital	0.8
Central Hospital	0.9
Medical College with Hospital	1.0

8. Communication	Weightage
Postal service Head Post Office (our building)	0.1
Post Office	0.2
Post Office with Phone	0.3
Telegraph Office	0.4
P&T Office	0.5
Post Office with Speed Post	0.6
Telephone Exchange	0.7
Courier Service	0.8
Mandal Post Office	0.9
District Post Office	1.0

9. Banking	Weightage
Private Finance	0.1
Self Help Groups (SHG)	0.2
Registered Financial Agency	0.3
Co-operative Society	0.4
Private Financial Institution	0.5
Nationalized Bank	0.6
Banks with ATM	0.7
Western Money Transfer (WMT)	0.8
Insurance	0.9
Bank with Insurance	1.0

Step (4): Hierarchy determination is based on centrality index (CI) and population served.

Step (5): Comparative Environmental system based on weightage techniques.

Step (6): Mapping techniques.

Settlement Geography

Urban Geography

Inter-city livability status in a developing economy: AGIS based case study (India): Shahab Fazal—

> "The present study attempts a geographic analysis for the concept of livability in a growing urban centre of inter-India." As the response of the population to the prevailing natural, social, cultural and economic environment and livability is behaviour related function of interaction between environment and personal characteristics. It is about the human requirement for social amenity, health and it includes both individual and community well-being."

Aligarh city

(1) Ward-wise status for built environment.
(2) Ward-wise status for social and cultural environment.
(3) Ward-wise status for economic environment.
(4) Ward-wise status for livability status/ranges very high, high, medium, low, very low, outside).

> "Like all ecosystems, the city is a system, having inputs of energy and materials and the environmental problems, economic lots as well as liability status are related to the growth of these inputs and managing increased outputs"

Literacy Trends – Urban-Land Differentials in India

—*Randhir Singh, Sarg Wan* and *Sangwar*

The present study relating to the description and interpretation of trends and regional aspects of urban rural differentials in literacy rates".

Step (1): To describe the literacy differentials by urban-rural residence.

Step (2): To analyse and interpret the trends in urban-rural differentials in literacy.

Step (3): To describe the spatial patterns of urban-rural differentials in literacy.

Methodology

Step (1)

Differentiate indices for literacy rates were computed on the basis of Sopher's Disparity Index (1974) as modified by Kundu and Rao (1982) is given below:

$$Ds = log\ (X_2/X_1) + Log\ (200-X_1)/(200-X_2)$$

Where $X_2^3 \times X_1$ and X_1 and X_2 are the literacy rates of alphas and non alphas respectively.

Step (2)

The Census Map of India (2001) has been adopted as the base map. The coefficients of correlation have been computed to determine the relationship between the dependent as well as the independent demographic and socio-economic variables. Finally, to ensure accuracy and to save time, the use of computer has been made.

Analysis: Literacy rates and differential indices by urban-rural residence, 2001

Persons/male/female – Total – literacy ratio %	urban-rural
Urban/rural	differential index

Step (3): Urban-Rural Disparity Ratio

Urban-rural literacy differential correlates, 2001

Variables	*Coefficient of Correlation*

(1) Rural literacy rate.
(2) Rural female literacy rate.

(3) Proportion of rural population in other services.
(4) Proportion of non-agricultural workers in rural areas.
(5) Level of urbanization.
(6) Rural Christian population.
(7) Proportion of SCs population to total population.
(8) Percentage of scheduled castes and scheduled tribes population of urban population.
(9) Proportion of scheduled castes and scheduled tribes population to total population.

Methodology: Coefficients of correlation are calculated by Spearmen's

Rank Differences Method (i.e.) $r = 1 - \dfrac{602}{N_3 - N} \quad \dfrac{r = 1 - 6\Sigma d^2}{N_3 - N}$

Urban Geography

Methodology (for studying the morphological evolution of a city)

Step (1): The objective of the present study is to evaluate the morphogenesis and spatial growth pattern of a city.

Step (2): To identify the morphology of the city as it has evolved and transformed the rough different periods of time and the proximate processes involved therein.

Step (3): **Source:** The information helps to study it from various sources including District Gazetteers of the study area (city), Census of India and other government documents and unpublished records of Municipal Committees.

Step (4): Maps have to be prepared to present different phases of development of the city. The maps like 'Morphological layout" phases of expansion, morphogenesis and 'Barrier's in expansion have to be prepared based on the information collected from various sources such as Town Planning department, urban development authority, district archives library and memory of senior citizens of different localities.

Step (5): **Innovative Items:** The 'shifting of Mean centres of settlement concentration' for different periods have to be plotted by computing mean location of settlements during *different periods on the graph and superimposing them on the map.*

Step (6): The line joining mean locations indicates the direction of shift of built up area in the city over a period of time.

The final map could be developed based on the Google Earth imageries

Settlement and land-use dynamism – A case study has been presented by M. Sakthival and important findings are as follows:

Important points are projected by objective and methodological patterns gains the attention of a prospective researcher.

The study of these tribes gains urgency under the circumstances uplifting the tribal communities and bringing them to the level of the national mainstream is the task of other developed communities around them. The author focused on dynamism of tribal settlements and land-use pattern should be taken up for their study.

Objective

(i) To study the land-use pattern and dynamism of Kalayar river basin, and

(ii) To analyse the spatial and temporal distribution of tribal during 1967 to 2005.

Methodology

(1) Survey of India topo-sheets 1:50,000 of 1967 are to be used for preparing the base map.

(2) The landforms are to be delineated from the relief map, based on relief and landform, the distribution of Kani settlement pattern and dynamism are to be carried out.

(3) *Result*: The present study analyse the spatial distribution of Kani settlements and their impact on hilly terrain of the western ghats region of Kanyakumari district, Tamil Nadu.

"The present study examines a comparatively unfavourable, high environment of steep hills with a developing economy and tries to analyse the pattern of change.

Time period: 1967-1997, the Survey of India maps on 1:50,000 were edited in 1967. Hence, 1967 has been chosen as the first time point, and 1997 has been chosen as the last point of time using *satellite imageries*".

Physiography (>1200 units): (1) The Northern hilly region, (2) The Eastern Southern Upland (600-1200 units), and (3) The isolated Southern hills and Southern plains (0-600 units). The general land use can be visualized.

Slope Analysis

(1) The slope analysis of river basin has been carried out, the broad division of slope units are identified (in metres) (0-5) nearly level; 5-10 (gentle slope), 10-15 (moderate slope), 15-20 (less steep slope), 20-25 (steep slope) and above 25 very steep slope.

Drainage Analysis

Power Projects

Land-use Map — 1967

(1) Dense forest,
(2) Moderately Dense forest,
(3) Forest plantation,
(4) Crop land,
(5) Plantation Agriculture,
(6) Open scrubs,
(7) Rock outcrop, and
(8) Waterbodies.

Settlement Geography (Micro Level Planning and Integrated Area Development Studies)

Objective

(1) To determine the hierarchy of urban settlements, by computing the composite centrality scores and then working out the hierarchy of urban settlements (using centrality scores); and
(2) To categorize the settlements into different hierarchical orders on the basis of computed centrality score.

Data Base

The study is mainly based on primary and secondary data collected from field work as well as from government offices of M.D.O/M.R.O. offices, Census of India (Population, Village Directory) and unpublished form of work firmly helps to alleviate the status of qualitative research besides the application of statistical and other quantitative methods.

Methodology

Hierarchy can be described as a first step to identify and recognize with differentiation of group of settlements. It depends on the size and kind of services offered by a settlement. It is a system of *ranking* method the towns and cities into successive groups based on the criteria like functions, size, etc.

Area of Study

An attempt is made to analyse the extent of availability of all infrastructural facilities in an area such as Tanuku Taluk, 1985, flood prone areas of Srikakulam District (2003) and Drought Prone area of Vizianagaram district, Andhra Pradesh (2007) were presented by Dr. Murthy for an intensive study see (in references).

Regional Development

Case

Regional Disparities in Levels of Human Resource Development — An Approach to Identification:

For the present investigation, the steps are to be followed:

(1) Mandal-wise or taluk-wise secondary data have to be collected from socio-economic review and District Statistical Abstract.
(2) Statistics other than primary source, the investigator should collect statistics from Zilla Parishad.

Methodology

(1) The collected data have to be processed and the method of

composite development have to be applied to find out the levels of development;

(2) One can identify distinct zones in levels in human resource development;

(3) High level of development could be observed in the Mandals/ Tahsils having high urbanization and high industrialization;

(4) Low level of development could be identified among the Mandals belonging to natural hazardous regions like flood prone/drought prone having mostly rural population and inadequate educational and medical facilities;

(5) Composite index of development method helps to find out the levels of development; and

(6) The results could be shown by Choropleth method; and

(7) For testing the hypothesis, '*t*' test could be employed to test the level of significance at 10 per cent. Composite index of development could be calculated by using the following indicators.

Indicators

1. Percentage of urban population to the total population;
2. Percentage of total literates to the total population;
3. Percentage of male literates to the total literates;
4. Percentage of female literates to the total literates;
5. Percentage of rural literates to the total literates;
6. Percentage of urban literates to the total literates;
7. Number of secondary schools per 10,000 population;
8. Number of degree colleges per lakh population;
9. Number of hospitals and dispensaries per lakh population;
10. Number of primary health centres and sub-centres per lakh population;
11. Number of hospital beds per 10,000 population; and
12. Percentage of workers engaged in secondary activities to the total working population.

Formula to be applied for calculating the levels of development

$$CD_1 = P_i/P_1 \times 100$$

CD_1 = Coefficient of development for variable '*i*'.

P_i = Percentage of variable '*i*' in the unit

P_1 = Mean percentage of variable '*i*' in the study region

$$C_1 D = CD_1^1 + CD_i^2 + CD_i^3 + CD_i^4 + \frac{n}{N}$$

C_1D = Composite index of Development
CD_i^1 = Coefficient of development for the index '*i*' in the enumerated unit.
N = Number of variables.

Hypothesis

(1) Higher the educational and medical facilities, higher the level of HRD.
(2) Higher the population of employment in secondary activities, higher the level of HRD.

Scale of Index value = (1) > 150 (2) 100 – 150 (3) 80 – 100 (4) < 80

7.5 REGIONAL PLANNING/DEVELOPMENT

A Micro Level Analysis – Role of Market Centres

Objective: To analyse the spatial distribution of market centres, their relationship to area, population, inhabited villages, net sown area and to study the role of market centres in levels of development.

Data Base and Methodology

Step (1): (1) Intensive fieldwork supplemented by secondary sources of data from socio-economic and District Statistical Abstract.
(2) The *Correlation Analysis* have to be applied to find out the relationship, whereas the *Composite Index* of development have to be applied in order to study the levels of development.

Step (2): Market Centres – Distributional Relationship

(1) Sl. No.	Per 100 Km²	Per 10,000 population

(2) Name of the Mandal $\bar{X}$ $\bar{X}$
(3) X – Average S.D S.D
(4) S.D. = Standard Deviation

Per 100 in habited villages — per 1000 hectares of net area sown

$\bar{X}$ $\bar{X}$

S.D. S.D.

Variables are to be related for Determining the Levels of Development

(1) Number of market centres per 100 KM^2 area;
(2) Number of market centres per 100 inhabited villages;
(3) Density of population per KM^2 area;
(4) Percentage of urban population to total population;
(5) Percentage of general literacy;
(6) Percentage of workers engaged in secondary and tertiary actual to total workers;
(7) Road density per 100 KM^2 area;
(8) Number of telephones per 10,000 population;
(9) Number of school facilities per 10,000 population;
(10) Number of colleges facilities per 10,000 population;
(11) Percentage of net sown area to geographical area;
(12) Percentage of net integrated area to net sown area;
(13) Number of banking facilities per 10,000 population;
(14) Number of cooperative credit societies per 10,000 population;
(15) Number of medical institutions per lakh population; and
(16) Beds available per lakh population.

Methodology: The coefficient of development of a mandal/tahsil in terms of single variable is calculated by the following equation

$$CD_i = P_i/P_1 \, X1a$$

Where CD_i = The coefficient of development for variable '*i*'.
P_i = Percentage of variable '*i*' in the unit.
P_1 = Mean percentage of variable '*i*' in the study region.

In order to develop a composite index, which reflects the composite effects of indicators enumerated above the following equation is to be adopted

$$C_iD = \frac{CD_i + CD_i^2 + CD_i^3 \ldots CD_i \times n}{N} \qquad [2]$$

Where C_iD = Composite Index of Development
N = Number of Variables

Mapping: Levels of development

Above 100 High
80-100 Moderate
Below 80 Low

7.5 Regional Studies

Identification of Regional Disparities in Levels of Development.

Objectives

(1) To examine the levels of regional development in social, economic and demographic indicators.
(2) To identify the socially and economically backward talukas/ mandals and categorize them into various levels of development.
(3) To find out various factors behind the emergence and growth of disparities in the study area.
(4) To find out the correlation between different factors which lead to regional disparities.
(5) To find out the overall development level.
Suggestions for future development.

Data Base and Methodology

Data Source

(1) State At A Glance - 2001;
(2) Census Handbooks;

(3) Directorate of Health and Family Welfare Service;
(4) Commissioner for Public Institution;
(5) Census of India – State; and
(6) District At A Glance.[8]

Generally, to derive composite index from a set of variables, a wide variety of *multivariate statistical techniques* are available. The choice of the most appropriate method depends upon the type of the problem, the nature of the data and objective of the analysis considering the above, *composite index method* have to be applied. For computing the composite development index at district level of any State, a composite picture has to be obtained by the following formula:

$$C_1 = \frac{\text{Observed Value}}{\text{Mean of the respective column}}$$

Where C_1 = Composite Index

The merit of composite index method is used to identify real variations among the different observations is the prime objective. The method of division by mean of the respective column (in the analysis) is to be preferred as it can get rid of the extreme values in the services. As far as the determination of weightages is concerned, no special weightages need to assign to any variable. It is to say that, equal weightage have to be assigned to all the selected variables to compute the development of the region. In this way, the removal of the biasness of scale, the transformed values of the given variables are added up lineally and the second values of composite index are classified into different categories. The categorized categories are (1) High; (2) Medium; and (3) Low developed regions, the standard deviation grouping technique is applied which is based *on the combined score* of 32 indicators.

Indicators

(i) *Social Indicators* - Education, Communication, Health facilities, and Recreation, etc. (X_1, X_2, H_3, H_4 - Sustn.]
(ii) Economic Indicators: (1) Regional Industries X_1, (2) Banking X_2, (3) Communication, (4) Transport, etc. [X_1, X_2, H_3, and

H_4], (5) Railways (X_5), (6) NAS (X_6), (7) Net Area (NA) (X_7), (8) Area Under HYV (X_8), (9) Fertilizer (X_9), (10) Vehicles (X_{10}).

(iii) *Demographic Indicators*: (1) Density of population (X_1), (2) Population growth (X_2), (3) Sex ratio (X_3), (4) Literates (X_4), (5) Urban literates (X_5), (6) Female literates (X_6), (7) Main working population (X_7), (8) Secondary workers (X_8), (9) Tertiary workers (X_9).

Methodology: Mean and standard deviation X and S.D. are to be computed

Clue: Observed values divided by mean of the respective column.

Regional Development Studies

There is every need to design a methodology for balanced regional development and levels of development.

Climate, soil and topographic conditions may predominantly dominate the physical conditions of the nation/state which in turn influence the life and economy of the people. Keeping in view of the above factors and it is apt to make an indepth and detailed survey and analysis of all the factors that influence the development of different physiographic region of the nation/state/or a study region. The nation or the state in general and real terms are the units and are only administrative regions with heterogeneous characteristics in climate, soil and topographic conditions, which makes the planner to prepare regional plans for development. The sub-regional level below the district level unit, for which the basic statistics are at present available.

Composite Index (C.I.) of development of different States/districts in a State have to be designed by the respective State governments based on selected socio-economic indicators. Further, (it helps) that the composite index of development helps really in identifying the developed as well as backward States/districts. The trend of growth and development could be attempted based on Standard Deviation Method (S.D.M.). It is noted that this analysis may help the centre/ State planning personnel/authority/government to identify the true and relative position of the State/district/mandal in various sectoral as well as overall development. This approach to some extent may help to tackle the problem of regional imbalances. Keeping in view of the problem in multiple dimensions, it appears that the most suitable strategy to overcome regional imbalances is to formulate techniques of treating regions is through regional plans for backward regions for example. Regional Plan for Telangana, on the basis of regional specialization and seek interaction between the rich resource region and the poor resource region by an interactive network of infrastructure facilities. The strategies should automatically attract not only the capital but also techniques of inflow into backward areas are to be designed. Besides this, profit maximization and welfare maximization principle could be achieved within a short period of time and also it is a strategy for socio-economic and infra development over a period of time.

'Planning' should not be in the name of planning but it should have to be a meaningful start with the utilization of local/regional physical resources a well as economic resources put in to achieve regional development through infra facilities, which facilitates the integrated development of the area/mandal/taluka/district/State or States of India.

The prime aim of the planner is to make a modest attempt to identify planning regions for regional development based on the aspects of criteria homogeneity, geographical contiguity, similarities of economic resources and problems to tackle the regional problems of regional unrest.

The selected indicators for assessment of levels of development should be able to present the true position of development in many respects, i.e., the quality of environment, settlement, infrastructure, Human Power (human resources, agriculture, industries and general economy. "The UNRISD have laid emphasis on, physical quality of life and identified methods to assess the `Physical Quality Life Index (PQLI).

Methodology

"The indicators have been standardized by adopting suitable procedure of classification with the help of mean and standard deviation. A composite picture has been obtained by using multivariate analysis techniques. With a view to assign weightages to the factor in the factor analysis, with due emphasis on the correlation matrix, a modified factor analysis has been employed by substituting the diagonals or correlation matrix by coefficients of multiple determination. The model employed here is $Yi = Bi\,Xi_1 + B_2Xi_2 + B_3X\,i_3 \ldots\ldots\ldots + Bn\,X\,in$

Where yi = the composite index for taluk i

Xi_1, Xi_2 Xi_3.....Xin etc., are the data inputs for the variables for the taluk i.

And B_1, B_2, B_3.....B_n etc., are the weightages obtained by diagnosing the modified correlation matrix and subjecting the diagonalised vector to the necessary weightages vector.

These Beta values have been subjected to the varimax rotation in order to obtain a perfection in communality and then the factor scores for each taluk has been estimated."[9]

Methodology

Factor Analysis (Theoretical Formulations).

Prime Aim: The prime aim objectives are; (i) To provide an insight into the trends in mandal disparities in the total framework of the macro/micro regional study; (ii) To construct composite indices of economic development, and (iii) To identify the backward, developed and highly developed mandals with respect to physical index, demographic index and infrastructural and overall development.

Methodology

To identify and recognize disparities of development, it is apt to apply factor analysis. The following variables are considered while adopting factor analysis technique.

- X_1 Total rainfall.
- X_2 Area Under Principal Crops (APC).
- X_3 Total Cropped Area (TCA).
- X_4 Area Under Principal Crops Irrigated (APCI).
- X_5 Rural Electric Service Connection for Agriculture (RESCA).
- X_6 Total Population (TP).
- X_7 Total Population S.Cs (TP SCs).
- X_8 Total Population S.Ts (TP STs).
- X_9 Educational Institutions (total number of institutions, TNEI).

X_{10} Postal Service (total number of Post Offices (TNP).
X_{11} Veterinary Services (total number of Veterinary Centres).
X_{12} Banking Services (total number of branches, TNBS).
X_{13} Medical Services (total number of hospitals, TNH)[10].

The limitation of these data was that the information on all 13 variables does not pertain to a single year. The data was collected from secondary sources from mandal revenue officer of 34 mandals relating to the selected variables. In this study the researcher is interested primarily to know the structural components of economy to construct composite indices of economic development, principal component method of factor analysis which has been used thus seeks to resolve a large number of measured variables in terms of relativity to few categories known as factors, each factor explains the maximum amount of variable percent. The greater advantage of this method is that it gives a mathematically unique solution. The factors so obtained are principal components (Kothari, 1990).

7.6 Modern Methods in Geographic Research *or* Role of Technology in Geographic Research

'Research' in Geography is of twofold. They are (1) Qualitative Methodology, and (2) Quantitative Methodology. The modern research methods are addition to the existing methods. They are (1) Scientific methods, and (2) Technological methods. The entire research activity has been broadly divided into four periods of study at the National level. They are (i) 1950-70, (ii) 1970-90, (iii) 1990-2010 and to the present day.

(1) 1950-70 Period

1950 to 1970, this is the early period in geographic research in India. During this period, much emphasis has been given to the core concepts and conceptualization of ideas and thought about the geographic phenomena and environment. It is primarily related to description, as the definition part of geography itself, i.e., Geo-Earth and Graphic Description. So, it is so happened that the geographic research is mainly based on primary collection of data, analysis and finally description of the existing process or pattern and their presentation. The questions of 'when' and 'where' about it are often put in by the geographers as well as non-geographers and the research descriptions are the answers for 'when' and 'where'. The first duration is 'when'. It refers to the historical event. It is to say/state that today's 'politics' is tomorrow's history. That is this particular incident which took place was an historical event which occurred on the space (place) or upon the earth. Further, we identify that geographers were primarily interested to identify and present the factual information and their description which took place in an elaborative and descriptive way. This leads to historical geography and political geography and its description. Indian geographers early contribution has been brought in this direction. Example Moonis Raza's research at AMU, Aligarh and later at JNU, New Delhi (14 *subas* of Muslim Rule in India – Medieval period).

If we examine the researches which were conducted in Agricultural geography and Industrial geography, they assume this type and direction. For examples, Irrigation landscape of Telangana. The entitled thesis was submitted by Dr. K.S. Reddy, Osmania University, Hyderabad, guided by Sri B.N. Chaturvedi and another one was industrial landscape of Telangana. Andhra Pradesh, was submitted by Dr. K.Vithal Reddy was guided by Sri B.N. Chaturvedi, Osmania University, Hyderabad – to Agra University, Agra and Vikram University, Ujjain, respectively during late sixties, i.e., 1969 and 1970.

It is to be noted that geographer's have specialized in historical geography, political geography, economic geography and the main focus was on 'Agriculture activity' was the major theme and equally on "Industrial activity" was quite thematic and descriptive. In the same period, we can also examine the thesis on "land utilization and occupational structure of people of Kolleru lake, Andhra Pradesh was submitted by Dr.Vidyanath, was guided by Prof. Manzoor Alam and second one was "coal mines of Kothagudem, Khammam district, Andhra Pradesh" thesis was submitted by Dr. Afzal Mohammed, Osmania University, Hyderabad guided by Prof. Manzoor Alam, Osmania University, Hyderabad, respectively.

The main aim was to present the site and situation, i.e., location and environment and the activity has been presented as on date.

(2) 1970-1990 Period

This period has been considered as young or formative stage and the research activity has been shifted towards quantitative oriented research with many variables of both qualitative and quantitative in nature. The geographic theories which were profounded and found to be more accurate and leads to scientific enquiry and judgment. The main focus was Urban geography/Agricultural geography and population and settlement geography. Western urban theoretical models and its applicability to Indian scenario was a theme to concentrate for current research topic. For example, Urban geography of Walled town: Warangal, Andhra Pradesh was carried out by Dr. Rama Mohan Rao for Ph.D. and the same was submitted under the supervision of Prof. Manzoor Alam, Osmania University, Hyderabad in 1974.

'Physical geography' and its thematic view with qualitative and quantitative methodology was developed by geographers.

The main focus was on physical geographic research in the sub-field of geomorphology, climatology and oceanography. Professor R.Vaidyanadhan's study is related to Cuddapah Basin and the main thrust was geological geomorphology. This study was conducted during (1953-63) at Andhra University, Waltair. Under his supervision, many research scholars have undertaken their studies leading to Ph.D. related to geographic geomorphology/landform and land-use/land cover/land-uses/land changes and morphometric (river basin) and the focus of study was primarily on field work supplemented by secondary data. The methodology applied in this research was focused with qualitative variables and quantitative variables, interpreting toposheet as basic map along with aerial photographic technique to delineate changes over a period of time. This researcher is able to identify and answer the questions like 'what' and 'how' and quantitative technology is the researcher's work.

'Climatology' and 'Oceanography' are the focus and represent the other sub-fields of physical geography. Eco-climatological studies and Oceanographic studies were mainly conducted in geography department of Andhra University, Visakhapatnam. Dr. Hema Malini, Dr. Sachidevi, Dr. Harikrishna and Dr. S. Dhanalakshmi have conducted their research studies with quantitative techniques. Koeppen and Thomthwente's climate classification and also water-balance techniques were used to identify the gaps and the results were presented. Agro climatological studies were undertaken by Dr. P. Viswanadhan and urban climatological studies were carried out Dr. Sachi Devi.

Simultaneously, the other subfields of geography and related research studies were also undertaken by Dr. Indira Devi's research which was on Urban geography of Visakhapatnam (1979) and another new branch was settlement geography and its focus was on Micro level planning and integrated area development.

'Micro level planning' – A Case Study of Nuzividu block, Andhra Pradesh has been selected for study by Mohan Kumar, K.V. and Integrated Area Development Planning – A Case Study of Tanuku Taluk, was by K.L.N. Murthy and these were supervised by Prof. P.V. Sarma, Economics Department, Andhra University, Visakhapatnam. The main focus of study was to identify the need to plan formulations and planning process should be focused at grass root level and also integrating the lower order settlements with the higher order central places and also identified the growth centres, service centres and central villages and this planning leads to spatial integration and functional integration of settlements finally arrived at integrated area development.

The main focus was central at the identification and recognition of theories (geographic) related to quantitative aspects and its applicability of particular themes whether they apply to the Indian conditions or not? The list is not exhaustive.

(3) 1990-2000 Period

The geographic research has been finding place in scientific organizations such as National Remote Sensing Agency/Organization/ Indian Space Research Organization/A.P. Remote Sensing Organization and also other State level organizations of Remote Sensing/A.P.Ground Water Board wherein geographers, efforts have been kept fulfilled. The focus of study was Remote Sensing Methodology/Aerial Photo Interpretation which has been applied in Geomorphology/Climatology/Oceanography and Change detection and identification studies. This period has been considered as mature stage in geographic research which has a wide appreciation and application of geographic tools with the existing science of geographical cartography with aerial photo interpretation techniques besides basic surveying led to quantitative geography.

The author is also interested to present the geographic research at the National level. The origin and development of geography at central universities and State universities (1) Aligarh Muslim University, Aligarh; (2) Banaras Hindu University, Varanasi; (3) Allahabad University, Allahabad; (4) Sagar University, Sagar; (5) Calcutta University, Calcutta; (6) Madras University, Madras;

(7) Bombay University, Bombay; (8) University of Pune, Pune; (9) Mysore University, Mysore (1960); (10) Osmania University, Hyderabad (1957); (11) Andhra University, Visakhapatnam (1972; and (12) Sri Venkateswara University, Tirupati (1972).

The researcher has carried out central theme at different universities which are focused and touched on multiple level aspects relating to major branches of geography with modern techniques such as Remote Sensing/GIS/GPS technologies.

(1) AMU – Aligarh : Agricultural Geography/Urban Geography and Settlement Geography, are major thematic research. Professor M. Shafi's interest was to find place on the agro-oriented studies relating to land utilization, nutrition and health, etc., later on, geographers have developed the other branches of geography such as urban and settlement geography. Nearly 180 Ph.D.s were produced as on date in Aligarh Muslim University, Aligarh.

(2) BHU, Varanasi : Urban Geography/Settlement Geography/ Regional Geography/GIS/RS/GPS are the major themes of geographic research. Professor R.L. Singh was interested to focus on Urban Geography and Settlement Geography, Kasinath Singh and R.B. Singh, Rana P.B. Singh's studies were related to Urban and Environmental Geography. It is observed that a recorded and indelible change in the research activity among the scholars are to concentrate on various branches/sub-fields of geography.

(3) Allahabad University: The geography department was started in year 1946 headed by R.N. Dubey, S.M. Ali and R.L. Diwedi. They have concentrated on Urban Geography of Allahabad and on various themes, Urban Geography has developed. Later on, S.S.Tiwari has developed school of agricultural geography as a major theme in geographical research. Savinda Singh has specialized on geographical geomorphology and various themes of geomorphology were developed besides mapping and cartography as a major theme.

(4) Sagar University, Sagar, Madhya Pradesh : The department of geography was started by S.M. Ali later on developed by N.P. Ayyar's study on agricultural geography of Narmada River was thematic research. Later on, Prof.V.C. Misra has developed the Department of Geography on various themes such as. Natural Rresource Study, Agricultural Geography/Urban Geography/ Settlement Geography and Geomorphology. Ph.D. theses were developed on the above themes with digital cartographic techniques which were appreciated widely.

(5) Calcutta University: The Department of Geography was started around 1947 and Professor S.P. Chatterjee headed the department. He was the authority on Geomorphology/Surveying and mapping and National Atlas and thematic mapping organization NATMO, Kolkata was started. This school of geography was developed in the major branches of geography such as physical geography with an emphasis on geomorphology river basin studies/ Agricultural geography/Settlement geography and Urban geography.

(6) University of Madras, Madras : The geography department was headed by Prof. G. Kuriyan and later on developed by Prof. A. Ramesh and Prof. P.S. Tiwari. The major thematic research that has been emerged out and fall into Agricultural geography/Urban geography, Settlement geography and the Modern geography has been developed by Prof. P.S. Tiwari using R.S./GIS/GPS, etc. Environmental geography is also an important theme besides medical geography in this school.

(7) Bombay University's Geography department dates back to 1950s and the major theme developed was Urban geography, later on geomorphology by Prof. K.R. Dixit, besides settlement geography and GIS/GPS/RS. Dr. K. Sita has developed Urban geography of Bombay during her regime.

(8) University of *Pune,* Pune: The geography department was started during 1950s and Professor C.D. Deshpande has developed Regional geography/Urban geography and later on geomorphology and modern geographical research methods like R.S./GIS and GPS.

(9) Mysore University, Mysore: Professor R.P. Misra was Lecturer in geography department under Prof. R.L. Singh during 1961 and later on, he had been to U.S.A. for pursuing Ph.D. and later on, Centre for Regional Studies was started during 1970s. The major themes are urban and regional planning with an emphasis to economic aspects of Regional Planning and the growth centre concept was central in this school.

(10) Osmania University, Hyderabad: The geography department was started by C.G. Smith in 1950 and later on developed by Prof. V.L.S. Prakasa Rao till 1968. The major themes are Urban Geography/Economic Geography. From 1968 to 1979, the major growth of geography was under Prof. S. Manzoor Alam. Geographers of later period have developed Geomorphology/Agricultural geography/fringe studies (urban fringe) with an emphasis on RS/GIS/ GPS and separately started functioning the department of Geo-informatics. Aerial photo interpretation secured a place in Urban studies/Settlement studies/Soil/Water/Air/Sound pollution studies are developed along with major geographic themes including Health geography/Nutritional geography 'Environmental' theme could find a new place in Modern geographic research.

(iv) Modern period 2001 to the present day: Ever changing ideas of man and available material technology has prompted the researcher to adopt new scientific approach to geographic research. During this period, innovative ideas which were emerged out from the researchers have taken shape into different sizes as scientific replica. The available methods of research were adopted as far as possible, i.e., such as the method of geographical cartography which has taken a new turn in the name of "Computer Aided Cartography (CAC) and digital cartography including digital photography and digital cartography besides photogrammetric methods and aerial photographs and its inter-disciplinary and applied research. Surveying has been taking a new turn in almost all Earth Sciences such as Geology, Geophysics, Geochemistry and Geomorphological methods to it are cartography and photogrammetric science. Geography, Cartography, Photography and allied sciences have taken new direction both in surveying methodology and techniques led to present the material with great precision through mapping techniques and technology. Modern technology helped in many ways to the researcher to grow to a maximum possible extent within a shortest period of time. The evaluation methods have assumed and replaced by new and modern technology such as (1) Remote Sensing technology (2) Geographic Information System (GIS), and (3) Geographic Positioning System (GPS), and (4) New technology used in interpretation and presentation of data and results in different cartographic media.

Many aspects of secured life to human is assured by government agencies in view of the existing insecurities. The major variables have been ever changing in tune with modern technology. The traditional geographic research has been replaced by Modern Science and technology, such as GPS and GIS, etc. Man who lives upon the Earth planet, under the sun, have been facing an endangering site and situation. 'Environment' has been playing an important and a predominant role, still men have many insecurities such as water, air, sound and soil are subjected to many pollutions with life harming pollutants. Traditional geographic research have a thrust on. 'Physics of the Earth', Economic Activity, Social action and interaction, cultural interaction lead to physical, economic, human and social geography. The new areas of research are listed below:

(1) Soil pollution, (2) Air pollution, (3) Sound pollution, (4) Ocean pollution, and (5) Environmental pollution.

Therefore, these branches are directly related to physical geography:

(1) Soil relates to pedology which leads to geomorphology,
(2) Air relates to atmosphere which leads to climatology, and
(3) Ocean pollution relates to Ocean surfaces which leads to Oceanography.

The new areas of research in economic geography are listed below:

(1) Agricultural geography – Nutritional geography and Health geography.
(2) Industrial geography – Geography of pollution – Geography of marketing.
(3) Human geography < population – urban population/rural population.
Settlement – urban settlement/rural settlement,
Geography of slums and Geography of migration.
(4) Regional geography – Morphology of a 'region'.

Regional Development – Structural Development – Equalities, Inequalities of development, spatial variations in resources – disparities of development and their levels – geographic causes hamper 'development'.

Planning for development — 'micro regions' for planning and development strategy for the overall development.

Methodology

The investigator should have to collect samples in order to identify the existing problems to the available resources (1) water resources, (2) Land-resource-use-changes, etc. Surface water and underground water, (3) Climate resources—Rainfall, temperature, precipitation, transportation and so on, (4) Ocean resources—surface informatics related to climate and beach characteristics and so on, (5) Health Care System (HCS), (5) Floods/Droughts/Tsunamis and Cyclones etc., (6) Watershed Management (Surface water and underground water) River Basin study, etc.

(1) *Water*: Water is a basic human need and a limited Natural Resource and a precious national asset. Human demand and misuse of water resources have continued to grow. Water security for human life, has become a matter of principal concern for sustainable development in the twentyfirst century.

Problem: Groundwater quality has its impact on human life. Good water saves human life. If fluorine concentration leads to fluoride and allied items leads to ill health.

Remedy: Collection of groundwater samples and they have to be analysed, the result of this study reveals a variation of fluoride content in water samples. An apparent observation of human being shows the symptoms of fluorosis, which kills the human life span.

Methods in Climatology

Scanty rainfall, low humidity, scarce vegetation, lack of water availability periods, extremes of temperature, intense of solar radiation and wind speed especially during summer and the characteristics of arid lands are important aspects of modern research. As a result of these factors, the potential evapo-transpiration (P.E.) in the arid region is very high in contrast to low precipitation. These elements attempts to analyse (Method) the water balance situation by applying *'Water Balance Methodology'* in the study area and it leads to vulnerability to desertification hazards. The hydrographic indications may reveal the need of water harvesting to support and subsist the water inadequacy of the weak natural ecosystem.

(1) Measurement of Water Scarcity: To Determine the Magnitude of Water Scarcity by Two Methods

(1) Thornthwaite (1948) method in which the magnitude of water scarcity has been determined on the basis of water level. High deficit means, higher the magnitude of water scarcity and lower deficit means lower level scarcity.

(2) This method is used for calculation of water scarcity index (2002), in which determinants for scarcity have to be selected.

Then by giving weightage, according to their importance, scarcity index have to be calculated. In this index, higher value represents lower level of water scarcity and lower value means higher magnitude of scarcity and finally magnitudes of scarcity have to be mapped and interpreted.

"Distantly, the district is infested by fluorosis disease. The data on dental care and skeletal fluorisis and content of fluoride in the ground water were collected from the records DMHO 2003. Analysis of the data indicated that fluorine has revealed that the fluorisis ranges between 3.0 and 5.5 parts per million in the drinking water. Connection of fluorosis is cases with fluorine content within the ground water indicated that high fluorine content of ground water associated with high incidence of the disease. The present study, the spatial patterns of fluorine content and fluorosis incidence were made".

(2) Epidemiological Diseases – Aids is associated with Human Immune Deficiency Virus (HIV).

Problem: Ill health caused due to HIV/AIDS.

Solution: Blood samples have to be collected and analysed. It should be done by any analyst.

Creation of Health Care Facilities (HCF): The intent of research is to attempt efficiency in health delivery through net working, upgrading facilities and efficient administration.

Health Care System, Health Care delivery politics, Health facilities development spatial availability should be made by government agency.

(3) Hospital Waste

"Waste management is one of the most essential services for maintaining the quality of life in urban areas and for ensuing better standards of health and sanitation – public health problems related with water and air pollution and solid infections have to be linked up to mismanagement of waste".

Wise Management: Proper steps to be taken up to restore good health for the population living in an environment. Distance decay, location-allocation and cost benefit analysis studies may help to identify the benefits.

Solution: To plan health facilities to the problem prone areas/ peoples. This may help to the establishment of Health Care Delivery Centre (HCDC) based on geographic variables (spatial studies).

Health Care System (HCS)

Problem & Studies

Criteria: (1) To study the existing situation of HCS; (2) To study the distributional pattern of HCS; (3) To find out relationship between socio-economic conditions and educational status of respondents; and (4) To suggest some remedial measures for improvement of the existing HCS.

(2) Noise Pollution

'Noise' is any sound that is not wanted, i.e., *unscientific* rather one of the common forms of atmosphere pollution. It is defined as unwanted high intensity sound without agreeable musical quality. The impact of Noise Pollution on human health, is to be examined and analyse the cause and consequences of noise pollution and their effects on human health.

Solution: To suggest measures and the same should have to be implemented by the government agency, should be referred to Pollution Control Board to take necessary action in this regard.

Action research is necessary in this regard.

Role of GIS, R.S. in Modern Geographic Research

Geographic Information System (GIS) and spatial data analysis are now allowing geographers and planners for more rigorous quantitative analysis of spatial phenomena. Remote Sensing (R.S.) and Geographic Information System (GIS) is the recently emerging area of research and job prospects for the geographers. The geo-informatics institute like Symbiosis Institute of Geo-informatics (SIG) aim to well train and empower the students with analytical decision making and management skills, besides enabling them to offer complex solutions in *infrastructure development, natural resource monitoring and urban development where in the data are analysed and used in areas like disaster management, environment, logistics and asset management to name a few.*

Computer Assisted Cartography (CAC) is emerging as another promising field of profession for progression to geographers. Terrain studies and beach characteristics like a number of bags, cover, inlets which not only provide good potential sites for studying the various coastal morphologic forms at present and process active therein but also serves as excellent facilities for shipping. Tidal flats, which get submerged during high tides daily and periodically.

Satellite imageries help to detect the current beach characteristics of bay areas, field measurement and mapping helps to reveal micro beach facets and features. Maps made using GIS reveal changes that have occurred over time also the impact of such changes on the livelihood of the local people and are noteworthy areas of research at present.

Work to be undertaken or Trust Area: Primary surveys should have to be conducted in order to understand coastal geomorphic process and offer solution to problem prone areas in India or at micro level.

Management of Watershed through Space Technology and GIS[11]

Watershed is a natural and hydrologic entity that cover a specific area in which rainfall flows to a defined drain channel stream system at any particular point. R.S. and GIS play an effective role in planning and integrated development of watershed. R.S. derived information can be merged and integrated with conventional data base. These diverse data can be put into computer system and may be modified, manipulated and converted into map format using suitable software.

Proposed to Study

(1) Medium resolution and multi spectral satellite data of khariff, rabi and zaid seasons and toposheet on 1/50,000 scale are to be used in generating information regarding drainage parameter, agricultural land-use and land cover and delineating watershed area.

The thematic information was generated using visual interpretation and digital image processing technique with the help of Erdas imagine and AVC GIS software. Image rectification, image to image registration, supervised and unsupervised classification of land-use and land cover have to be taken up with the help of Erdas

Imagine. Various individual layer such as drainage, rail and road network, irrigation structure and land-use and land cover map have to be generated with the help of AVC GIS. In showing the relief features in and scope map of the study area are to be prepared with the help of AVC GIS. The statistical data related to land-use and land cover, population and settlements are entered in attribute data set.

Source: A case study of Mansata Watershed of Sova on Tehsil of Allahabad District. Sunil Kumar Tripathi, Archana Tripathi, Allahabad is an exemplary study of the latest branch of research.

Role of GIS and Remote Sensing in Disaster Management

Generally, area and people suffer very often from various natural disaster, i.e., drought, flood, earthquake, landslide, and forest fire, etc. These natural disasters are of (1) Geographical origin such as earthquakes, volcanic eruptions, landslides, (2) climatic origin such as drought, flood and forest fire which strike causing a devastating impact on human life, economy and environment.

It is possible to be away to some extent from it.

(1) Minimize the potential issues by developing early working strategies;
(2) Prepare and implement developmental plans to provide residence to such disasters;
(3) Mobilize resources including communication and tele-medicinal services; and
(4) To help in rehabilitation and post-disaster reconstruction. Space technology plays crucial role in efficient mitigation of disasters. Communication satellites help in disaster warning, relief mobilization and tele-medicinal support.

Earth observation satellites provide required data base for pre-disaster prepared programmes, disaster response, monitoring activities, and post-disaster damage assessment, reconstruction and rehabilitation

GIS helps to evolve a suitable strategy for disaster preparedness and operational framework for their monitoring, assessment and mitigation, identifies gap areas and recommends appropriate strategies for disaster mitigation.

Watershed Management

Remote Sensing (R.S.) and Geographic Information System (GIS) play very significant role in watershed management by providing multi data and reliable data/images, techniques of interpretation, analyzing complex data sets and their mapping as well as storing these informations for future investigation. Different aspects of watershed management such as physiography, drainage, soil, vegetation, hydro geo-morphological information, land-uses/land cover, land degradation sites/erosion intensity and socio-cultural phenomena and natural implements related to plan implementation can easily be assessed and mapped by using RS and GIS techniques. "The data base has been generated by collecting information from different sources such as SOI, NATMO and NRSA, Hyderabad/Remote Sensing data of IRS, IPZISS III, and IRS PZ LISS III (1:50,000) IRS IDLISS III digital data (1:2,50,000).

Forest Resources Management using RS

"Forest is a vital resource for life, economy and habitat of human beings. It provides fuel, fodders timber, medicinal herbals, raw materials for industries and numerous ortho products needed in every walk of life. It helps in controlling cloud drifting, floods, droughts, soil erosion, spring, land slides, desertification, temperature evaporation and increasing rainfall, maintaining oxygen balance and in retaining water holding capacity."

The development of RS, GPS and GIS technologies have greatly helped independent investigations of forest cover change for effective planning. Forest Survey of India (FSI) has planned to monitor and map correcting forest cover on a biennial basis. Consequently, FSI has developed a methodology based on high resolution satellite data for mapping and stratification of forest, leading to an improved sampling design for field inventors. The forest cover assessment uses Digital image processing software and involves the acquisition of satellite data, geometric rectification of raw data, mosaicing of rectified scenes and classification of forest cover using NDVI.

"To have a data base system on various planning issues and problems needed for the multifaceted developmental programmes".

Uses GIS, RIS in coastal erosion, on set of summer monsoon over India, impact of climate change in water resources planning and management, morphometric evaluation of drainage basins, spatial pattern of slopes, uses: geographical boundaries could be demarcated by using toposheet and ecological models could be developed (for conversion and restoration of ecosystem of the lake).

Planning for sustainable ground water management, drainage morphometry of river basins, climate change and flooding "tracing out the application of recent techniques of RS, GIS system Flood Hazard Mapping (FHM) should be in a more rational way of flood management retain them formulating structural measures along the dynamic rivers. The regulation of flood hazard areas coupled with enactment and information of flood hazard zoning could prevent damages to life and property in a long way."

Typomorphic

This study attempts to –

(1) To identify soil erosion area;
(2) To explain the formation of rills and gullies;
(3) To assess the improper land-use which causes accelerated soil erosion; and
(4) To recommend some suggestions for better management strategy.

The interaction of 'high land-low land' borne factors:

(1) To investigate topographical, hydrological, morphological and anthropological causes of floods;
(2) To investigate the flood impacts, mitigation and management, flood warning dissemination and response; and
(3) To evaluate planning issues associated with the city and its varied communities with respect to sustainability of the unit".

Environmental Impact Assessment (EIA) is used to identify the environmental and social impacts of a project prior to decision making and also understand the people's opinion, perception as well as mis-apprehension about the project (socio-economic survey) alarming adverse effects could paralyze the good health of human being in due course of time. It is an integrated approach to sustain and save our ecosystem from different levels of environmental pollution.

Due to large scale withdrawal of ground water from acquifers, the pilzometric level has gone down to 5-10 m and if the situation continues, there would be a threat of vacuum of water in the hand granite rock of the region. Artificial recharge is a technique of augmentation of water resources.

NOTES

1. Based on Dasgupta and Misra (1978).
2. http.//www.irimet.net, 2) http.//www.mpo
3. T.V. Krishna Rao, M. Harikrishna, K.V. Mohan Kumar, "Development of GIS data-base for the Assessment of Environmental Impact....." (pp. 148-153).
4. Murty, K.L.N. "Applied Geo-informatics" Chapter 8, pp. 123-134, Mangal Deep Publications, Jaipur, 362 018.
5. Ram Mohan Rao, and Simhadri, S (1984): *Method of Measuring Agricultural Efficiency*, Readings in Agricultural Resource Development (ed.), Hyderabad Publishers.
6. Krishan Chand Ramotra: Geographical Analysis of Population Pressure on Land in Maharashtra, India", *IJRS*, Vol.37, No.1, 2005, pp.138-144.
7. Curve fitting analysis has been carried out by Dr. C.S.S.R.L., Hanumantha Rao, Lecturer in Statistics, M.R. (A) College, Vizianagaram, Andhra Pradesh.
8. R.Y. Singh: *Geography of Settlements* Published by Rawat Publications, Jaipur (India), P.VII-2010.
9. Particular case study relates to Bellary District of Karnataka State by L.T. Nayak and D.S. Narayankar may be referred to Vol. 41, No. 1, 2009, *Indian Journal of Regional Science*, Kolkata.
10. For detailed Analysis and text, please refer to Dr. T.N.A. Rao, *Balanced Regional Development – Levels of Development of Karnataka*, Published by Academy of Regional Science and Humanities, (ACARSH) Belgaum, Karnataka, India.
11. For detailed study please refer to K.L.N. Murty, *Factor Analysis (Theoretical Formulations and Conceptual Strategy)* Chapter 8, pp.123-134.
12. A case study of Mansala Watershed of Sova on Tehsil of Allahahad District Sunil Kumar Tripathi, Archana Tripathi, Allahabad.

REFERENCES

A. Kamalakar Reddy and Vijay Bhole, *Geographical Perspectives in Environmental Issues.*

Annals NAGI, Kavita Misra, V.K.Kumara, S.P.Mishra, *Morphometric Analysis*, (pp. 11-31).

Annals of NAGI, Vol. XXVIII, No. 2, December, 2008.

Annals of the NAGI, Vol. XXXI, No. 1, June 2011.

Annals, *NAGI*, Vol. XXIX, No. 1, June, 2009.

Annals, *NAGI*, Vol. XXIX, No. 1, June, 2009.

Annals, NAGI, Vol. XXVII, No. 1, June, 2007.

Annals, NAGI, Vol.XXIX, No.2, December, 2009.

Annals, *NAGI*, Volume XXIX, No. 2, December, 2009.

Annals, of the NAGI, Vol.XXVIII, No. 2, December, 2008 (pp.18-31).

Annals, Vol. XXX, N.1, June, 2010.

Annals, Vol. XXXI, No. 1, June, 2011 (pp. 81-92).

Anuradha Saha and Rashmi Ranjana, *Air Pollution and Health Hazard – A Case Study of Patna*, pp. 60-74.

Applied Geo-informatics, Published by Mangal Deep Publishers, Jaipur (India), 2007.

C.R. Pathak Ed., Essays in Regional Science, Regional Development and Planning, Vol. 1, Silver Jubilee Issue.

C.R., Pathak Editor (1995): Essays on Regional Science, Regional Development and Planning, Volume-1, Silver Jubilee Issue, (pp. 299-311).

Johnston, R.J. (1978) Multivariate Statistical Analysis in Geography, Longman, London, pp. 127-182.

K. Lingebiel, A.A. and Montgommery, P.H. (1961): *Land Capability Classification, USDA, Handbook No. 210*, Ministry of Agriculture, Fisheries and Food (1968) – Agricultural Classification Map of England and Wales, London.

Kavita Mishra, V.K. Kumara and S.P. Mishra, "Morphometric Analysis and Prioritization of Micro-sub Watersheds" (pp.11-31), *Annals*, Vol. XXIX, No. 2, December, 2009.

King, L.J. (1969) Statistical Analysis in Geography, Prentice Hall, Englewood Cliffs, New Jersey.

M. Sakthivel, *Kanisettlements and Land Use Dynamism*, pp. 85-95.

M.V. Reddy: *Agriculture and Agro-Based Industry in Chittoor District*, (pp. 63-66), published with the financial assistance from University Grants Commission, New Delhi.

Manoj Kumar and M.S. Jaglaw: *Morphological Evolution of Hisar City*, (75-90).

Meetai M.M. (2000): "Rural settlement system of Imphal East and West District" An Unpublished Ph.D. Thesis, Manipur University

Moumita Talukdar and *et al.*, *Land-Use Changes in Tinibukia District, Assam.*

Muliman, A (2002): "Spatial Analysis of Traditional and non-traditional market place exchange: A case study of Raichur District", Unpublished Ph.D. Thesis, Karnataka University, Dharwad

Murthy K.L.N. (1986): *Planning for Integrated Area Development – A Case Study of Tanuku Taluk*, Ph.D. Thesis (Published), B.R.Publishing Corporation, New Delhi.

Murthy, K.L.N. (2003): *Ecology, Environment and Development*, Mangal Deep Publication, Jaipur, (Rajasthan) India.

Murthy, K.L.N. (2007): *Applied Geo-Informatics*, Mangal Deep Publication, Jaipur (Rajasthan), India.

Murthy, K.L.N. "Land Use Planning Model for Srikakulam and Environs: An Urban Dynamic Plan in a Rural Setting". – 27th International Geographical Congress, Washington, 1992.

Murthy, K.L.N.: Applied Geo-informatics – A Case Study from Drought Prone Areas, pp.94-103, Mangal Deep Publishers, Jaipur (India), 2007.

Murthy, K.L.N.: Ecology, Environment and Development – A case study from Flood Prone Areas, pp. 127-153.

Murthy, K.L.N.: *Planning for Integrated Area Development – A Case Study from Andhra Pradesh*, pp. 193-196.

Narpat Singh Rathod, "Application of Multi Temporal Satellite Data for the Study of Water Resource Problems in Udaipur Basin, South Rajasthan", pp. 72-88.

R.B. Singh and Ajay Kumar Gurjar, *Climate Change Vulnerability of Agriculture*, *Annals*, Vol. No. 31, No. 1, June 2001.

R.R.L. Kantam Prof., M. Krishna Reddy Prof. and others, *Statistical Methods and Inference by Telugu Academy.*

R.S.Sangwan and S.Sangwan, *Literacy in 20th Century and Trends in its Urban-Rural Differentials in India*, pp. 79-94.

S. Ramesh, *Land use Suburbia,* pp. 93-99.

S.B. Gaikwad, C.T. Power: "National Pattern of Sugar Industry in Maharashtra, C.G.A" and N.S.Gaikwad, in *Indian Journal of Regional Science*, Vol. 39, No. 1, 2007, pp. 150-153.

S.C. Gupta (ed.), V.K. Kapoor, *Fundamental Mathematical Statistics.*

Shahab Fazal, "Land Transformation in Relation to Distance in Developing Economy", *Indian Journal of Regional Science*, Vol. 38, No. 1, 2006, pp. 91-104.

Statistical Abstract, India, 2001. Agriculture Pattern of Land Utilization ('000 hectares).

Surendra Singh and V.S.Chauhan, "Resource Utilization and Potentials" (pp. 81-82)" in *Regionalization for Rural Development and Planning*, Independence Publishers, Meerut-2005001 (U.P.), India.

V.S. Bhadrapur and S.S. Naregal: *Agricultural Efficiency Regions and Regional Disparities: A Comparative Study of Bellary District*, pp. 277-287.

8

Statistical Geography

Problems on Arithmetic and Geometric Series

1. Write down the next ten terms of the following series

(i) 5, 7.5, 10, ...
(ii) 12, 8, 4, ...
(iii) $x - y, x, x + y, \ldots$

(i) 5, 7.5, 10, 12.5, 15, 17.5, 20, 22.5, 25, 27.5, 30, 32.5, 35
(ii) 12, 8, 4, 0, –4, –8, –12, –16, –20, –24, –28, –32, –36
(iii) $x - y, x, x + y, x + 2y, x + 3y, x + 4y, x + 5y, x + 6y, x + 7y, x + 8y, x + 9y, x + 10y$

2. Write down the few terms of the series formed by $4n$–1. Find the n-th term of the series whose first term is $(x + 2)$ and the common difference is 3

If n = 1, then, value = 3
If n = 2, then, value = 7
If n = 3, then, value = 11
If n = 4, then, value = 15
If n = 5, then, value = 19
If n = 6, then, value = 23
If n = 7, then, value = 27
If n = 8, then, value = 31
If n = 9, then, value = 35

n–th term of an arithmetic progression $a + (n - 1)d$ where 'a' is the first term of the series and d is the common difference.

In the given problem $a = x + 2$ and $d = 4$
n–th term = $a + (n-1)d = (x + 2)4$

3. Find the 15th term of the series

0.6, 0.72, 0.864, …

In the Geometric progression n–th term is given by where a is the first term and r is the common ratio

In the given problem 0.6, 0.72, 0.864

$A = 0.6$, $r = 1.2$,

The 15th term = $a \times r$ 15.1
= 0.6 × (1.2)14
= log 0.6 +14 × log 1.2
= 7.709

4. A parent places in the saving bank Rs. 5 on his son's first birthday, Rs. 10 on the second, Rs. 15 on the third and so on. How much will be saved up when the boy becomes a major?

In the given problem, a parent places in saving bank Rs. 5.

On his first birthday, Rs.10 on his second birthday and 15 on third and so on upto he becomes major attaining 18 years.

we can put money he saves in the series

5, 10, 15 … 85

The sum of the first 'n' natural numbers is given by $n(n + 1)/2$

In the present problem $n = 18$

So applying formula = 5[18(18+1)]/2
= 5(9×19) = 855

Rs. 855 will be saved by the time the boy becomes major.

5. Find the tenth term of the series

(i) 5, 10, 20.....
(ii) 1.1, 1.21, 1.331,.....
(iii) – 0.5, 0.15, – 0.045,.....

In the problem all the series exhibit geometric progression

In the G.P n–th term is given by $a \times r^{n-1}$, where a is the first term and n is the common ratio.

(i) 5, 10, 20, ...
Where $a = 5, r = 2$
10th term is $a \times r^{(n-1)} = 5(2)^9 = 2560$

(ii) 1.1, 1.21, 1.331, ...
Where $a = 1.1, r = 1.1$
10th term is $a \times r^{(n-1)} = 1.1(1.1)^9 = 2.535$

(iii) –0.5, 0.15, –0.045, ...
Where $a = -0.5, r = -0.3$
10th term is $a \times r^{(n-1)} = (-0.5)(-0.3)^9 = 0.0000098415$

6. Find the arithmetic and geometric mean in each of the following cases:

(i) 3 and 5
(ii) 4.2 and 3.6

Sol: Arthmetic mean

(i) 3 and 5
Arithmetic mean = 3+5/2 = 8/2 = 4
Geometric mean

(i) 3 and 5 $= \sqrt{(3\times5)}$
= Average of log 3 and 5
= (0.47771 + 0.6990)
= 1.1761
= 0.58805

Anti log = 3.873
So G.P of 3 and 5 = 3.873

7. Find the sum of the First fifteen terms of the series 4+5+6.25+ ...

Sol.: The series is 4, +5, +6.25 ...
In the term in a G.P is given by $a \times r^{n-1}$
In here a = First term, r = common difference
Here $q = 4$, $r = 5/4 = 1.25$
In the Geometric series the sum of the n term is given by $a(1-r^n)/ (1-r)$ or $a(r^{n-1})/ (r-1)$

In here a = First term and r = common ratio
The sum of 15 terms of the series is

$$= 4 \times (1.25)^{15} - 1 /(1.25-1)$$
$$= 4 \times (1.25)^{15}/0.25$$
$$= 432.56$$

Topic: Calculation of Mean, Mode and Median

1. Given the following frequency distribution, calculate the mean

Class	*Frequency*	*Mid values*	*Frequency*	$X_i f_i$
12.5 – 17.5	2	15	2	30
17.5 – 22.5	22	20	22	440
22.5 – 27.5	19	25	19	475
27.5 – 32.5	14	30	14	420
32.5 – 37.5	3	35	3	105
37.5 – 42.5	4	40	4	160
42.5 – 47.5	6	45	6	270
47.5 – 52.5	1	50	1	50
52.5 – 57.5	1	55	1	55
			72	2005

Arithmetic mean $\Sigma^n X_i f_i / \Sigma f_i$

$$\Sigma X_i f_i = 2005$$
$$\Sigma f_i = 72 = 2205/72 = 27.84$$

2. Find the median and mode for the following distribution

Marks	*No. of Students*
0 – 4	10
4 -8	12
8 - 12	18
12 – 16	8
16 -20	4
20 -24	5

Marks	*Mid value*	*Frequency*	*Cumulative above*	*Frequency below*
0 – 4	2	10	10	57
4 – 8	6	12	22	47
8 – 12	10	18	40	35
12 – 16	14	8	48	17
16 – 20	18	4	52	9
20 – 24	22	5	57	5
		57		

$$\text{Medium} = \frac{l + (n/2 - c)}{f} i$$

Where l is the lower boundary of the medium class
f is the frequency of the medium class
c is the cumulative frequency up to and not including medium class
N is the total frequency
I is the class interval

$l = 8\ N = 57,\ \frac{N}{2} = \frac{57}{2} = 28.5,\ c = 22,\ \text{f} = 18,\ \text{i} - 4$

$$l + \frac{(n/2 - c)}{f} i$$

$$8 + \frac{(28.5 - 22)}{18} \times 4$$

$$8 + \frac{(6.5)}{18} \times 4 = 9.44$$

$$\text{Mode } i = \frac{fm - f_1}{2fm - f_1 - f_2} \times h$$

Where $i =$ is lower limit of the model class
fm is the maximum frequency
F_1, f_2 are the frequencies of the classes

Preceding and following the model class

h is the width of the class interval
$i = 8\ fm = 18\ f_1 = 12\ f_2 = 8,\ h = 4$

$$t + \frac{f_m - f_1}{2f_m - f_1 - f_2} \times h$$

$$8 + \frac{18 - 12}{2 \times 18 - 12 - 8} \times 4$$

$$8 + \frac{6}{16} \times 4 = 9.5$$

3. Topic: Mean Deviation, Standard Deviation and Coefficient of Variation

1. Calculate the Mean Deviation from the Mean of the following distribution

Marks	0-10	10-20	20-30	30-40	40-50
No. of students	5	8	15	16	6

Ans: Assumed Mean = 25 (A)

Class	*Mid Values*	*f*	*cf*	*d* $\frac{(x-25)}{10}$	*f.d*	*[x-27]*	*f [x-27]*
0-10	5	5	5	–2	–10	22	110
10-20	15	8	5	–1	–8	12	96
20-30	25	15	13	0	0	2	30
30-40	35	16	28	1	16	8	128
40-50	45	6	44	2	12	18	108
							472

Mean = $A + id = 25 + 10 \times 18/50 = 25 + 2 = 27$

$$\text{Median} = L_1 + \frac{(n/2 - c)}{F} i$$

20 + 10 × 12/15 = 28

Mode = Mean – 3 (Mean – Median)
= 27 – 3 (27 – 28)
= 27 – 81 + 84 = 30

Mean Deviation from mean = Σ *f*|x – 27 |Σ*f*
= 472/50 = 9.44

Mid values	*Frequency*	*(x – 28)*	*Fx (x – 28)*	*(x – 30)*	*Fx (x-30)*
5	5	23	115	25	125
15	8	13	104	15	120
25	15	3	45	5	75
35	16	7	112	5	80
45	6	17	102	15	90
			478		**490**

Mean Deviation from Median = 478/50 = 9.56
Median Deviation from mode = 490/50 = 9.80

The mean deviation about the median should be less than the mean deviation about the median. But in this problem we got the mean deviation about the median is greater than the mean deviation about the mean.

4. Topic : Calculation of Correlation Coefficient

x	*y*
1	9
2	8
3	10
4	12
5	11
6	13
7	14
8	16
9	15

Obtain an estimate of *Y* which should correspond on the average to *X* = 6.2

X	1	2	3	4	5	6	7	8	9	= 45*X* = 45/9 = 5
Y	9	8	10	12	11	13	14	16	15	= 108*Y* = 108/9 = 12

$$X - Y = 9\ 16\ 30\ 48\ 55\ 78\ 98\ 128\ 135 = 597$$

$$(X - \bar{X}) = -4\ -2\ -1\ 0\ 1\ 2\ 3\ 4$$

$$(Y - \bar{Y}) = -3\ -4\ -2\ -1\ 0\ 1\ 2\ 3\ 4$$

$$(X - \bar{X})^2 = 16\ 9\ 4\ 1\ 0\ 1\ 4\ 9\ 16 = 60$$

$$(Y - \bar{Y})^2 = 9\ 16\ 4\ 0\ 1\ 1\ 4\ 9\ 16 = 60$$

$$\bar{X} = 45/9 = 5$$

$$\bar{Y} = 108 / 9 = 12 \quad X\beta\eta = \sqrt{\frac{1}{n}\Sigma(x - x)^2}$$

$$Y = \sqrt{\frac{1}{9} \times 60} = \sqrt{\frac{20}{3}}$$

$$Y = 108 / 9 = 12\ X = \sqrt{\frac{1}{n}\Sigma(y - y)^2}$$

$$Y = \sqrt{\frac{1}{9} \times 60} = \sqrt{\frac{20}{3}}$$

$$X = y = \sqrt{\frac{20}{3}}$$

$$\text{Cov}(x, y) = \frac{1}{n}\Sigma Xy - \bar{x}.\bar{y}$$

$$= 1/9 \times 597 - 5 \times 12$$

$$= 199/3 - 60 = 19/3$$

$$R = \frac{\text{cov}(x, y)}{xy} = \frac{19/3}{\sqrt{\frac{20}{3} \cdot \frac{20}{3}}} = \frac{19/3}{20/3} = \frac{19}{20} = 0.95$$

$$Y - \bar{y} = r = x\frac{(x-\bar{x})}{\sigma / x}$$

$$\begin{aligned} Y &= \bar{Y} + r\,(x - x) \\ &= 12 + 0.95\,(6.2 - 5.0) \\ &= 12 + 0.95 \times 1.2 \\ &= 12 + 1.40 \\ &= 13.14 \end{aligned}$$

The estimate of *Y* correspond on the average to *X* is 6.2 is 13.14

Calculate the Standard Deviation from the following data

Class interval	Frequency (*f*)
5.5 – 6.5	3
6.5 – 7.5	6
7.5 – 8.5	9
8.5 – 9.5	13
9.5 – 10.5	8
10.5 – 11.5	5
11.5 – 12.5	4

Solution:

Class interval (i)	*Mid point (m)*	*Fre-quency (f)*	$M \times f$	*Devia-tion from mean (a)*	*Devia-tion squared up (d)*	*Squared deviation x frequency* $(fx - a^2)$
5.5 – 6.5	6	3	18	–3	9	27
6.5 – 7.5	7	6	42	–2	4	24
7.5 – 8.5	8	9	72	–1	2	9
8.5 – 9.5	9	13	117	0	0	0
9.5 – 10.5	10	8	80	+1	1	8
10.5 – 11.5	11	5	55	+2	4	20
11.5 – 12.5	12	4	48	+3	9	36
		$N = 42$	$\Sigma mf =$ 332			$\Sigma fd^2 = 332$

$$\text{Arithmetic Mean} = \frac{\Sigma mf}{n} = \frac{332}{48} = 9$$

$$\text{Standard Deviation} = \sqrt{\left(\Sigma fd^2 / n\right)}$$

$$= \sqrt{\left(\frac{124}{48}\right)}$$

$$= \sqrt{2.58} = 1.6$$

Karl Pearsons Method of Coefficient of Correlation

Karl Pearson has introduced a formula for computation of coefficient of correlation between another two series by dividing the product of deviation from intimate average by the product of two Standard Deviation to the total number of pair unit. This is given as

$$r = \frac{\text{S}dxdy - n}{nx\phi xx\phi y}$$

where r = coefficient of correlation, dx = deviation of item values of x series from its arithmetic average, dy = deviation of item values of series from its arithmetic average. $D \times dy$ = product + $dx \times dy$, i.e., $dx \times dy$ = product Qx = Standard Deviation of y series.

Shortcut Method

$$r = \frac{\Sigma dxdy - n\left(\frac{\Sigma dx}{n}\right)\left(\frac{\Sigma dy}{n}\right)}{nx\phi xx\phi y} \quad \ldots \text{(i)}$$

$$r = \Sigma dxdy - n\ \frac{(\Sigma dx)}{n} - \frac{(\Sigma dy)}{n}$$

$$nx = \frac{\sqrt{\Sigma dx^2} - dx\left(\Sigma^2\right)}{n} \times \frac{\sqrt{\frac{\Sigma dy^2}{n} (\Sigma dy)^2}}{n}$$

Example: Given are the weight of 10 fathers and their sons. The problem is to find out the correlation between the two.

Weight of in fathers in Kg's	63	64	65	67	68	69	70	71	71
Weight of the sons in Kg's.	65	63	63	65	67	67	68	71	68

Weight of the fathers	*Deviation from Asset*	*(dx^2) Square of Deviation*	*Weight of the son (y)*	*Deviation of 67 (dy) from*	*Square of (dy)*	*Product of (x^2) (dx^2) and Dx, dx, dy*
63	–5	25	65	–2	4	10
64	–4	16	63	–4	16	16
65	–3	9	63	–4	16	12
67	–1	1	65	–2	4	2
68	0	0	67	0	0	0
69	+1	1	67	0	0	0
69	+1	1	68	+1	1	1
70	+2	4	71	+4	16	8
71	+3	9	68	+1	1	3
71	+3	9	69	+2	4	6
	+3	75		–4	62	58

$$\frac{\Sigma dxdy - n\left(\frac{\Sigma dx}{n}\right)\left(\frac{\Sigma d4}{n}\right)}{nx\left(\frac{\Sigma dx}{2}\right)-\left(\frac{\Sigma dx}{n}\right)^2 \times \left(\frac{\Sigma d4}{n}\right)-\left(\frac{\Sigma d4}{n}\right)^2}$$

$$\frac{58-10\times-\frac{3}{0}\times 4/10}{10\times\sqrt{\frac{75}{10}-\left(-\frac{3}{10}\right)\times\frac{62}{10}-\left(\frac{4}{10}\right)^2}}$$

$$=\frac{58-10\times-.03\times 4/10}{10\times\sqrt{7.5-0.09\times 6.2-0.16}}$$

$$= \frac{56.8}{10 \times \sqrt{7.41 \times 6.2 - 0.4}}$$

$$= \frac{56.8}{10 \times 2.72 \times 2.46}$$

$$= \frac{56.8}{66.9} = 0.85$$

Correlation coefficient (Grouped data)

Given the following data, obtain the correlation between X and Y

X / *Y*	30-35	35-40	40-45	45-50	50-55	55-60
80-90	2	3	2			
90-100		2	5	4	2	
100-110		4	8	5	1	
110-120			2	3	1	1
120-130				2	1	1

X	32.5	37.5	42.5	47.5	52.5	57.5	*f*	*f*v	*f*v²	*f*u	*f*uv
	-3	-2	-1	0	1	2					
Y 85-*Z*	2	3	2				7	-14	28	-14	28
95-1		2	5	4	2		13	-13	13	7	7
105 0		4	8	5	1		18			-15	
115 1			2	3	1	1	7	7	7	1	1
125 2				2	1	1	4	8	16	3	6
F	2	9	17	14	5	2	N≥49	Σ-12	64	-32	42
*f*u	6	-18	-17		5	4	-32				
*f*u²	18	36	17		5	8	84				
*f*v	-4	-8	7	3	1	3	-12				
*f*uv	12	16	7		1	6	42				

Correlation coefficient between *X* and *Y*

We have to convert them into *f*, *u* and *v*

Here $u = \dfrac{x-475}{5}$; $v = \dfrac{Y\text{-}105}{10}$

Correlation coefficient ηXY = η and v

Correlation coefficient

$$\eta xy = \eta \text{ and } v = \frac{\Sigma fuv - \frac{1}{n}[\Sigma fu.\Sigma fv]}{\sqrt{\Sigma fu^2 - \frac{1}{n}(\Sigma fu)^2} . \sqrt{\Sigma fv^2 - \frac{1}{n}(\Sigma fv)^2}}$$

$$= 42 - \frac{1}{49}[-32X - 12]$$

$$= \frac{42 - 1/49[-32X - 12]}{\sqrt{84 - 1/49(-32)^2}} . \frac{1}{\sqrt{64 - 1/49(-12)^2}}$$

$$= \frac{42.7 \times 83.7}{\sqrt{84 - 20.9}\sqrt{64 - 29.4}} = \frac{34.169}{\sqrt{63.1}.\sqrt{61.06}}$$

$$= 0.55.4$$

Topic: Trend by the method of moving average

The production of a commodity in millions of tonnes is given below:

For the years

Year	*Production*	*Year*	*Production*
2001	52.4	2006	38.3
2002	38.2	2007	47.0
2003	45.0	2008	31.8
2004	46.5	2009	42.5
2005	39.7	2010	41.9

Work out a five year moving average and a 4 year central moving average. Draw the original series and these trend series in a single group.

Trend by the method of moving averages

Year	*Produc-tion*	*3 year Moving Average*	*4 year Moving Average*	*4 year Central Moving Average*
2001	52.4			
2002	38.2		45.525	
2003	45.2	44.36	42.35	43.935
2004	46.5	41.54	42.37	42.36
2005	39.7	43.30	42.875	42.62
2006	38.3	40.66	39.2	41.08
2007	47.0	39.86	39.9	39.55
2008	31.8	40.30	40.8	40.35
2009	42.5			
2010	41.9			

Demographic Constants

(1) Calculate the crude, age specific and standardized death rates of the local population from the following data and compare them with the crude death rate of the standard population

Age Group (1)	*Standard Population* (2)	*Number of Death* (3)	*Local Population* (4)	*Number of Death* (5)
0-10	700	20	500	18
10-20	1100	7	1600	8
20-60	3100	26	2500	26
60-100	500	22	800	23
Total	**5400**	**75**	**5400**	**75**

(*Contd.*)

(*Contd.*)

Age Group (1)	*Age specific death rate local population* (6)	*Age specific death rate total population* (7)
0-10	36	28.6
10-20	5	6.36
20-60	10.4	8.4
60-100	28.75	4.4
Total	—	—

The crude death rate of the standard population =

$$\frac{\text{Number of deaths}}{\text{Total population}} = \frac{75 \times 1000}{5400} = \frac{375}{27}$$

The crude death rate of the local population = $\frac{75 \times 100}{5400} = 13.9$

Standard death rate for local population = $\frac{\Sigma R_i P_i}{\Sigma P_i}$

Where R_i = Age specific death rate of the local population
P_i = Standard population of that age group

∴ Standard death rate for the local population

$$= \frac{(36 \times 700) + 5\,(1100) + 10.4\,(3100) + 28.75\,(500)}{700 + 1100 + 3100 + 500}$$

$$= \frac{77315}{5400} = 14.32$$

The crude death rate of the local population is same as that of the standard population. The standardized death rate of the local population is greater than the crude death rate of the population.

(2) The following is the distribution of male and female population according to age groups:

Age group	*Population (in 1000)*	
	Male Population	*Female Population*
0-14	94	90
15-29	50	57
30-49	52	47
50-69	23	20
70 and above	5	6

Calculate from the above data :

1. Sex ratio.
2. Ratio of women in the reproductive age group to the total women and to the total population.

Answer:

Age group	*Male*	*Female*	*Sex ratio Female/Male×1000*
0-14	94	90	957.4
15-29	50	57	114.0
30-49	52	47	903.8
50-69	23	20	869.6
70 and above	5	6	1200
Total	**224**	**220**	

Total women in the reproductive age group in 15-49 = 104000
Total women = 2,20,000

(i) Ratio of women in the reproductive age group

$$\text{In total women} = \frac{104000}{2,20,000} \times 1000 = 473.0$$

(ii) Ratio of women in the reproductive age group in the total

$$\text{population} = \frac{104000 \times 1000}{444000} = 23423$$

EXERCISE

Draw Bar diagram for the given data

Jute production for the year 1986-1987

S. No.	*Name of State*	*Arid Irrigated to total percent*	*Production (% to total Production)*
1.	West Bengal	64.0	67.3
2.	Bihar	17.7	14.5
3.	Assam	12.0	11.9
4.	Orissa	4.6	4.6

Ans: Hint (Draw vertical bars)

Draw pie diagram for the following data:

Size	*Cities*	*Percent*		*Population (000's)*		*Urban Population to Total Population*	
		1991	2001	1991	2001	1991	2001
I	39 (28 USA)	15.02	22.54	11913	15440	66.88	75.3
II	43 (BUA)	15.96	24.85	2245	2859	12.66	13.94
III	46 (UAS)	42.73	26.59	2945	1711	16.53	8.35
IV	23	18.31	13.29	587	346	3.30	1.69
V	20	6.57	11.56	114	139	0.64	0.68
VI	2	1.14	1.15	9	7	0.05	0.04
Total	**173**	**100**	**100**	**17,813**	**20503**	**100**	**100**

Represent the following data in suitable diagram:

Index of per capita income (1993-94 Base)

S. No.	*Region*	*1974-79*	*1995-2000*
1.	Northern coastal region	79.2	98.0
2.	Southern coastal region	112.6	105.8
3.	Coastal Andhra	103.6	103.7
4.	Rayalaseema	96.9	92.1
5.	Northern Telangana	92.1	91.2
6.	Southern Telangana (without Hyderabad and Rangareddy)	80.7	95.4
7.	Hyderabad & Rangareddy (without Telangana)	87.5	92.9
8.	Hyderabad & Rangareddy	140.6	122.3
9.	Telangana	97.2	99.6
10.	Andhra Pradesh	100	100

Hint: Horizontal bars for the years 1974-79 and 1995-2000 respectively.

Major Food Crop Productivity for the year 2002-2003

S. No.	*District*	*Food crops*	*Paddy*	*Oil seeds*	*Sugar-cane*
1.	Srikakulam	301.7	269.2	26.0	207.3
2.	Vijayanagaram	160.3	117	34	579.5
3.	Visakhapatnam	137.7	64.5	16.0	579.5
4.	Northern coastal region	599.7	450.7	76	1366.3
5.	East Godavari	1103	1043	1.0	1269.6
6.	West Godavari	1304.3	1165.7	7.6	2916.0
7.	Krishna	959.4	766.9	8.0	1535.3
8.	Guntur	861	654.2	10	114.9
9.	Prakasam	307.3	151.3	27.0	24.4

Hint: Food crop productivity of the same year and different crops such as food crops, paddy, oilseeds and sugarcane be presented both in pie diagram/vertical/horizontal bar diagrams.

REFERENCE

Singh, K. *Technique and Method of Social Survey, Research and Statistics*, pp. 346-349, Prakashan Kendra, Lucknow, 1983.

8.1

Research Paper

Population of India : A Geographical Appraisal and a Statistical Account*

Dr. C.S.R.L. Hanumantha Rao[1] *and Dr. K.L. Narasimha Murthy*[2]

- Synopsis
- Abstract
- Introduction
- Earlier Studies
- Hypothesis
- Objectives
- Geographical Appraisal
- Statistical Account/Assessment
- Future Trends
- Conclusion

Abstract: An attempt is made in this paper to present the past, present and future trends of population growth of India as it helps the planners to plan the infrastructural facilities to the growing needs of increasing population besides providing food security to the existing

---

* Paper communicated to 34th Indian Geography Congress, November 2-4, 2012, Patna University, Patna, (Bihar), India.

1. Lecturer in Statistics, Department of Statistics, MRA College, Vizianagaram, Andhra Pradesh, India.

2. Principal (Retd.), Government UG and PG College (Men), Srikakulam – 537001, Andhra Pradesh, India.

population "Population Potential" is considered always as a source of energy and a resource to the country to achieve development over a period of time. As time goes on, the space will be reduced to human beings because of the dearth of resources, the ever growing population may have face the retarded development condition and, hence, the population growth should have to be reduced alarmingly so as to maintain equity and equality as a none to maintain the desired number optimum level as far as possible curve fitting methodology, and moving average method has been applied to estimate the future population Suggestions for reducing the population is given at the end to safeguard India from the optimum level of population and it has its impact and imprint not only on the production but also on many aspects of human life.

Introduction

Population growth / trend is understood as a process of concentration and involves two aspects (1) Multiplication of Population over a period of time; (2) The increase in the size of population in different census years, i.e., growth of population in various census years. The growth or trend meaning is somewhat larger in terms of it as a process of change relating to demographic, economic, social, cultural and geographical dimensions. There is a strong relationship between growth and trend such as linear, quadratic and exponential. The interpretation of this change is highly remarkable and it is difficult to find exactness on this point. The variance in this particular aspect could be observed even at the macro level. In India also, the exact differences are witnessed by various census years in different States and Union territories is important to identify and recognize changes.

The objective of this paper is to analyse the nature of trends by statistical methods such as curve fitting and also identify the 95 to 100 per cent change over a period of 25 to 30 years (Over three census years) which is almost visible and compared the figure, to a tune of 95 per cent change of population (geographical interpretation) correlates to statistical assessment or statistical account and India has been experiencing the population explosion. Accordingy, the aspects of trend/change of population at the national level and changes have been considered even in the intracensus years. Calculation of inter-census population could give the clue further to understand such trend/change in population. The increasing population have its impact on multiple variables/sector such as food supply, nutrition, health, education, transport and communication and finally on infrastructural development at the national level.

Data and Methodology

An analysis has been made at the national level by using the census data. Here, this study too is based on census figures (various census years) obtained from census of India reports and different volumes.

The analysis made here follows statistical methods besides a discussion made on national net increase in population in particular the trend/growth in between the two census years, are analysed. It is observed that the number of States have been changing in India like all other States. Therefore, census-wise temporal changes are analysed by using arithmetic calculation/statistical account.

Earlier Studies

Zimmerman clearly emphasized that man is the most dynamic agent of production and the beneficiary of the entire process of a resource development and utilization. He is the pivot around which resource processes move. Resources presuppose a person and they are the expression of human appraisal (Zimmerman, 1951, p. 7) it is, therefore, said that the concept of natural resource is largely desired from our own societies and ceaseless attempt at finding new and more intensive uses for the raw materials of the nature (Alexander Spochr, 1965, p. 110). Archana Bhargava,[1] has attempted to present the analysis relating to interrelation between population – its number, and qualities at one hand, and the natural resources on the other besides to analyse the spatial patterns of the above attributes as far as possible. The author has analysed the distribution of population, density patterns, (by the land and man ratio) and spatial patterns of the arithmetic density relating to high density of population, low density of population and moderate density of population besides presenting spatial patterns of the nutritional density relating to a case study, i.e., Sagar division, M.P. Krishan Chand Ramotra[2], S.K. Vadiyar and D.H. Pawar have presented, "the geographical analysis of population pressure on land in Maharashtra" wherein they have presented the population pressure on land by considering a few important Indices like growth rates of population and foodgrains, cultivated land and foodgrains availability per head to visualize the magnitude of the problem. They have further investigated that population pressure on land has been increasing continuously since 1971 and concluded that almost the entire State of Maharashtra is under Population pressure.

K.L.N. Murthy[3] has presented the demographic patterns and identified that human resource is a factor of prime importance in the economic development of any region and man's role in the society as well as resource utilization is an important factor and provides the required labour and skill to plan and perform the process. The interplay of man and resources determine the stages of development. The population pressure or concentration enables the author to divide the whole area (study region) into seven types of concentration. They are designated as (1) Concentration 'A' type; (2) Concentration 'B' type; (3) Concentration 'C" type; (4) Concentration 'D' type (5) Concentration 'E' type; (6) Concentration 'F' type; and (7) Concentration 'G' type and the changing pattern of range trends.

Narasimha Murthy, K.L.[4] has presented the *population Growth A Geographical interpretation* and the study of the patterns of growth in India in population by applying statistical methods at various stages, they are (a) Newton's Forward Interpolation formula; (b) Newton's Backward Interpolation; (c) curve fitting; and (d) Regression equation with the help of the past census figures, enable the author to estimate the future growth patterns of India's population, density of population arrived at, that an increase in the physical inputs will definitely and periodically effect the space and yields spatial problem.

S.B. Singh[5] have presented, *Impact of Urbanization Upon the Environment of Varanasi City*" and opined that the natural increase of population of the urban residents but also the unimpeded influx of migrants from suburbs and rural areas, which has goaded the crisis and further identified that, the ever increasing population of the urban areas leads to the over exploitation of resources as the per capita demand of resources increases. Finally, he has presented to rationally apprehend the relationship between man, environment and urbanization and environmental deterioration and observed that, "the pattern of spontaneous and haphazard development because much aggrevated after independence, as the city was flooded with the Hindu Migrants from Pakistan". The author has presented the population change in Varanasi City, 1901 – 2001, the variables are (1) Population, and (2) Increase or decrease and growth rate (percentage).

Distribution and Density of Population in India, 1981

"There are qualitative changes in the distributional pattern of population not only in the world and also in India. In this context, that the studies concerning population distribution and density assume significance for less developed countries.

Ahmad's (1941) and Geddes (1942) studies of the pre-independence era notwithstanding Chatterjee (1962) using the 1951 census of India, stands but as the solitary attempt during the post-independence period dealing with this aspect of Indian population at the national level. However, some sporadic attempts were made at regional level including those of Kuriyan (1938), Verma (1956), Chatterjee (1961), Sinha (1958), Krishan (1968), Ghosh (1970), Prakash (1970) and Mehta (1973) which covered only Kerala, Punjab, West Bengal, Orissa, Bihar, Uttar Pradesh and Rajasthan. Ghosal and Chandna (1979) while taking stock of research work done in population geography during 1969-72, stressed the need for examining patterns of population distribution and density in the country by using the latest census data. Chandna has presented the distribution of population through the concentration index technique and explains further that the concentration index map has thus been preferred to a dot map because of non availability of actual location of over half a million rural settlements in the country[6].

An area of influence together with the city in town is known as "city Region or unland". The area of influence of an urban centre can be determined and delimited as the basis of distance decay curves of selected variables. The influence of city is gradational in nature. As the distance from the city centre increases, its dominance wanes and also the influence of some other urban centres becomes greater (K. Markandey, 1986). They have presented fitted equations of distance decay functions of second degree of polynomial functions, correlation coefficient (r), r_2, t values of beta coefficients (t_1, t_2) F-value and the value of x at $dy/dx = 0$ (in km). In a paper entitled, "City – Unland Relationship of Class I and II Town of Rayalaseema" by C. Sujatamma, M.V. Reddy and N.B.K. Reddy[7].

Surendra Singh and Chauhan, V.S.[8] have presented the trend of population growth, a reliable estimate in this regard have made for the last seven decades (1901-71), during which period the population of Rohilkhand has grown from 5. I million to 10.4 million, recording an increase of 68.8 per cent. Growth of population during the corresponding period in U.P. was 81.7 per cent, which was comparatively higher than that of Rohilkhand. Further, stressed that total population growth of Rohilkhand was slightly increased during the decade of 1901-11, but perceptible decrease is noted between 1911 and 1921. Later the growth was, more or less, constant. The trends of population growth (Simpson & Kafka, 1965, 235 – 71) for the last seventy years may be shown as under x:

$$Yc = 6.88 + 1.89x$$

The trend of population growth is calculated by straight line equation, i.e., $yc = a + bx$ (where 'a' indicates mean population and 'b' indicates the growth gradient).

"Rural population growth dynamics in Chandigarh periphery zone" was presented by Pawan Kumar Sharma.[9] Sharma have calculated the compound annual growth rate, the initial year has been considered as the base for calculating the compound annual growth rates, and has presented the compound annual growth rate of Rural Population by population size category of villages (1971 – 2001), population groups on x-axis and per cent on y-axis in order to identify the compound annual growth rate into groups through a graph. The variables are considered such as (i) distance from Chandigarh (in kilometres) 0-4, 4-8, 8-12; and 12-16 (2 percentage of rural area (1971-72) 3 percentage of villages (1971-72) and percentage share of village population (1971-72) and percentage share of Rural population by sub-zones (1971-2001. Finally, the author has presented the percentage share of area and population according to distance from Chandigarh.

Ritu Parna Ghosh and Lakshi Siva Ramakrishnan[10] have presented Urban Infrastructure in Durgapur Municipal Corporation – A Spatial Analysis in a paper and attributed that, Urbanization in developing countries is characterized by rapid growth of population in the major urban centres as they provide economic opportunities and most of the amenities that are required for living quality human life. However urban infrastructure is not uniformly developed within an urban area and further, it is found to be well developed in the central a core area of a city and progressively declines with distance from the city centre. The selected Indices indicates some trends as postulated by Burgress model besides Lorenz curve and Ginnis method of coefficient have also been worked out to assess spatial inequality in distribution of urban amenities and they have identified the aspects from the above study are (1) functionality, (2) spatial diversity and (3) core periphery relation, and centrality.

Jabir Hasan Khan, Shamshad and Tarique Hassan[11] have presented thematically in their paper, "Determinants of Rural to Urban Migration of Male Population in India". Centered the aim to analyse the socio-economic causes of rural to urban migration of male population in the age group of 15 to 59 years and whose duration of residence was one to four years, among the States and Union Territories of India.

They have identified that, "the ratio of total rural urban migrants, whose duration of residence was one to four years, was recorded 21.43 per cent to the general migrants of the same duration, corresponding figures for male and female migrants were 31.07 per cent and 16.23 per cent respectively.

Keeping in view of the aforementioned literature, the uncovered area of literature aspects so far are considered and covered in this paper with a humble attempt.

4. Hypothesis

It is hypothesized that the population of India has been on an increasing scale to a tune of not less than 90-95 per cent added to the total population at the national level over a period of 2 to 3 census periods, if there are no checks over population.

5. Objective

The author's main objective is to identify the growth patterns / trends of India's population ranges from 1901 to 2011 later years are and crude arithmetic calculations, which have any correlation to statistical accounting procedure. This could be achieved through the statistical methods such as (1) curve fitting; and (2) moving average, and the above methods may also help to the analysis to a tune of 95 per cent satisfaction and one should have to appraise of the geographical interpretation besides statistical validity by using the Census of India figures to detect the growth pattern and find the trend values, which are presented in four cases. Statistical calculation is presented in Table 4, 5, 6, 7 and 8.

Table 1: Change in population, 1951-2011

S. No.	Census year	Population in crores	Time lag.	Value	Population	Change in population
1.	1951	36.11	t_0	—		
2.	1960	43.92	t_1	$T_1–t_0$	43.92–36.11	7.81
3.	1971	54.82	t_2	$T_2–t_1$	54.82–43.92	10.90
4.	1981	68.33	t_3	$T_3–t_2$	68.33–54.82	13.41
5.	1991	84.64	t_4	$T_4–t_3$	84.64–68.33	16.31
6.	2001	102.87	t_5	$T_5–t_4$	102.87–84.64	18.13
7.	2011	121.02	t_6	$T_6–t_5$	121.02–102.87	19.15

$$\bar{x} = \frac{\Sigma xi}{N}$$

$$\text{Change} = t_1 + t_2 + t_3 + t_4 + t_5 + t_6 = \frac{85.71}{6} = 14.285 \text{ crores}$$

Table 2: Second degree difference

S. No.	Year	Population in Crores	Change	2nd Degree Difference
1.	1951	36.11	—	—
2.	1961	43.92	7.81	3.09
3.	1971	54.82	10.90	—
4.	1981	68.33	13.41	2.51
5.	1991	84.64	16.31	2.90
6.	2001	102.87	18.13	1.82
7.	2011	121.02	19.15	1.02

Table 3: *Method* 3: Identification of trend in population (by the method of moving average)

S. No.	Census year	Population	2nd year moving total	2 year moving average figure
1.	1951	36.11	—	—
2.	1961	43.92	80.03	40.01
3.	1971	54.82	98.74	49.37
4.	1981	68.33	123.15	61.05
5.	1991	84.64	152.97	76.48
6.	2001	102.87	187.51	93.50
7.	2011	121.02	223.89	112.05

6. Geographical Appraisal

It can be seen from Table 1, after the independence of India, the census figures have been taken for seven census years, i.e., 1951 as base year, i.e., to 1961, 1971, 1981, 1991, 2001 and 2011. The population in crores (growth) is indicated in column (4) and the formulae to detect change is given in col (5) and the change in population has been stated in 7th column. If we observe the change in population, the ever increasing trend is noticed. From 7.81 (1951) to 19.15 (2011).

Change of population on an average is to a tune of 14.285 crores.

Identification of net change in population is presented in Table 2. The 4th column of Table 2, indicates the change is population the change over change, i.e., second degree of difference could also be seen in the last column of the table.

It can be seen from Table 3, the identification of trend in population by applying the method of moving average. 2 year moving totals and 2 year moving average figures are presented in column 4 and 5. A percetible change of population (trend) could be seen and it would be more visible if we plot the values of actual figures, and upon it 2 year moving total, and two year moving average figure gives the actual volume of change (trend) in population over a period of time.

7. Population Increase by Estimation (By Arithmetic Method)

Case (i): If we hypothesize that "Indian Population will be doubled within 30 years period."

According to 1971 Census, the population of India was

$$54.82 \text{ crores } x_2 = 109.64$$

According to case (i) hypotheses that the population
For 2001 will be to a tune of 109.64
Instead it is grown to a tune of (–) 102.87
The difference as per estimation is to a tune of 6.77 crores

109.64 – 6.77

$$\frac{100 \times 6.77}{109.64} = \frac{677}{109.64} = 6.174 = 100 - 6.174 = 93.83 \text{ per cent}$$

Case (ii) according to 1961 census, the population of India was 43.92 crores

As per the hypothesis that Indian population will be doubled within 30 years, i.e., 1961 + 30 = 1991

Will be 43.92 × 2 = 87.84 crores

The 1991 population of India was 84.64 crores

The difference between characteristic is 87.84 – 84.64 = 3.2 crores

$$87.84 - 3.2$$

$$\frac{100 \times 3.2}{87.84} = \frac{320}{87.84} = 3.64 \;\; = 3.64$$

$$= 100 - 3.64$$

$$= 96.36 \text{ per cent}$$

Case (iii): According to 1981, the Indian population was 68.33 crores. It is hypothesised that Indian population will be doubled within 40 years (68.32 × 2 = 136.64 crores)

Statistical calculation for the 2021 population will be 130.59 (by curve fitting formula)

The difference is (136.64 – 130.59) = 6.05 crores

$$130.59 - 6.05$$

$$\frac{100 \times 6.05}{130.59} = 4.63 \text{ per cent}$$

$$100 - 4.63 = 95.37 \text{ per cent}$$

Case (iv): According to 1991, population was 84.64 crores. If we hypothesize that Indian population will be doubled within 30 years, i.e.,

$$1991 + 30 = 2021$$

$$= 86.64 \times 2 = 173.22$$

(ii) If we hypothesize that India population will be doubled within 40 years, i.e., 1981 + 40 = 2021 = 68.33 × 2 = 136.66 crores.

(iii) The growth rate from 2001 is continued till 2021.

The Indian population will be

2001 – 102.87 = 18.15

2011 – 121.02

The difference from 2011 to 2001 is (121.02 – 102.87) = 18.15

If the growth is estimated to a tune by half

$$\frac{18.15}{2} = 9.75$$

$$= 121.02 + 9.70$$

$$130.72 \text{ crores}$$

Formulae

(i) $(t_1 - t_0)$

(ii) $t_n = (t_1 - t_0)/2$

By applying the (ii) formula

$$(121.12 - 102.87)/2$$

$$= 18.5/2 = 9.71 \text{ crores}$$

$$= 121.02 + 9.71$$

$$= 130.73 \text{ crores}$$

As per arithmetic calculation the population for 2021 would be 130.73 crores

As per the curve fitting (statistical account) the population for 2021 would be 130.59 crores.

The difference of population estimation as per arithmetic formulae and statistical calculation is only 0.14 (130.73 – 130.59)

$$130.73 - 0.14 = 100 \times 0.14 / 130.73 = 1\%$$

Statistical calculation in assessment of Indian population + Population growth: India 1901 to 2011.

The population of India at the turn of 20th century, was only around 238.4 million. This has increased by more than four times in a period of the hundred and ten years to reach 1210 million in 2011. The population of India grew by one and half times in the first half of the 20th century while in the later half, it recorded a phenomenal three fold increase. Statement 1, presents the population of India as recorded in each decade of census since 1901. Some other indicators of growth rate such as decadal growth rate, change in decadal growth, average annual exponential growth rate and progressive growth rate over 1901 during each decade have also been presented in this statement.

Table 4: Population and its Growth, India: 1901 – 2011

Census year	*Population*	*Decadal growth absolute*	*Per-cent*	*Change in decadal of growth absolute*	*per cent*	*Average annual expo-nential growth rate (per-cent)*	*Progre-ssive growth rate over 1901 (per cent)*
1901	238396327						
1911	252093390	13697063	5.75			0.56	5.75
1921	251221213	–772177	0.31	–14469240	6.05	0.03	5.42
1931	278977238	27656025	11.00	28428202	11.31	1.04	17.02
1941	318660580	39683343	14.22	12027317	3.22	1.33	33.67
1951	361088090	42427510	13.31	2744168	0.91	1.25	51.47
1961	439234771	78146681	21.64	35719171	8.33	1.96	84.25
1971	548159652	108924881	24.80	30778200	3.16	2.20	129.94
1981	683329097	135169445	24.66	26244564	0.14	2.22	186.64
1991	846421039	163091942	23.87	27922497	17.12	2.16	255.05
2001	1028737436	182316397	21.54	19224455	10.54	1.97	331.52
2011	1210193422	181455986	17.64	–860411	0.47	1.64	407.64

Source: Provisional population totals – India 41.

1. Population

Table 5: Variation in population 1901 to 2001

S. No.	*Year*	*Population*	*Variation (+) or (–)*		*Growth Rate*	*Density of Population (Per sqr. Km)*
			Persons	Percentage		
1.	1901	958778	–	–	–	147
2.	1911	1007388	48610	5.07	5.07	154
3.	1921	1018268	10880	1.08	1.08	156
4.	1931	1119891	101623	9.98	9.98	171
5.	1941	1235800	115909	10.35	10.35	189
6.	1951	1347764	111964	9.06	9.06	206
7.	1961	1415961	68197	5.06	5.06	217
8.	1971	1589558	173597	12.26	12.26	243
9.	1981	1804196	214638	13.50	13.50	276
10.	1991	2110943	306747	17.00	17.00	323
11.	2001	2249254	138311	6.55	6.55	344

Source: Director of Census Operation, Hyderabad.

Table 6: Ranking of States and Union Territories by Population: 2001 and 2011

Rank in 2011	*India/State/ Union Territory*	*Population 2011*	*Per cent to total population of India*		*Rank in 2001*
1	2	3	4	5	6
	INDIA	1210193422	100.00	100.00	
1.	Uttar Pradesh	199581477	16.49	16.16	1
2.	Maharashtra	112372972	9.29	9.42	2
3.	Bihar	103804637	8.58	8.07	3
4.	West Bengal	91347736	7.55	7.79	4
5.	Andhra Pradesh	84665533	7.00	7.41	5
6.	Madhya Pradesh	72597565	6.00	5.87	7
7.	Tamilnadu	72138958	5.96	6.07	6
8.	Rajasthan	68621012	5.67	5.49	8
9.	Karnataka	61130704	5.05	5.14	9
10.	Gujarat	60383628	4.99	4.93	10
11.	Orissa	41947358	3.47	3.58	11
12.	Kerala	33387677	2.76	3.10	12
13.	Jharkhand	32966238	2.72	2.62	13
14.	Assam	31169272	2.58	2.59	14
15.	Punjab	27704236	2.29	2.37	15
16.	Chhattisgarh	25540196	2.11	2.03	17
17.	Haryana	25353081	2.09	2.06	16
18.	NCT of Delhi	16753235	1.38	1.35	18
19.	Jammu & Kashmir	12548926	1.04	0.99	19
20.	Uttrakhand	10116752	0.84	0.83	20
21.	Himachal Pradesh	6856509	0.57	0.59	21
22.	Triupra	3671032	0.30	0.31	22
23.	Meghalaya	2964007	0.24	0.23	23
24.	Manipur	2721756	0.22	0.22	24
25.	Nagland	1980602	0.16	0.19	25
26.	Goa	1457723	0.12	0.13	26
27.	Arunachal Pradesh	1382611	0.11	0.11	27
28.	Pondicherry#	1244464	0.10	0.09	28
29.	Mizoram	1091014	0.09	0.09	30
30.	Chandigarh#	1054686	0.09	0.09	29
31.	Sikkim	607688	0.05	0.05	31
32.	Andaman & Nicobar Islands#	379944	0.03	0.03	32
33.	Dadra & Nagar Haveli #	342853	0.03	0.02	33
34.	Daman & Diu #	242911	0.02	0.02	34
35.	Lakshadweep #	64429	0.01	0.01	35

Notes: See notes 4 & 5 below statement 2.

Provisional Population Totals – INDIA 47.

Table 7: State-wise Distribution of Population and Total Workers 1971 – 2001 (excluding Assam, Jammu & Kashmir and Mizoram)

S. No.	State	Total Popn (1971)	Total Popn (1981)	Total Popn (1991)	Total Popn (2001)
1.	A & N Islands	115133	188741	280661	356152
2.	Andhra Pradesh	43502708	53549673	66508008	76210007
3.	Arunachal Pradesh	467511	631839	864558	1097968
4.	Bihar (including Jharkhand)	56353369	69914734	86374465	109944338
5.	Chandigarh	257251	451610	642015	900635
6.	D & Nagar Haveli	74170	103676	138477	220490
7.	Delhi	4065698	6220406	9420644	13850507
8.	Goa (including Daman & Diu)	857771	1086730	1271379	1505872
9.	Gujarat	26697475	34085799	41309582	50671017
10.	Haryana	10036808	12922618	16463648	21144564
11.	Himachal Pradesh	3460434	4280818	5170877	6077900
12.	Karnataka	29299014	37135714	44977201	52850562
13.	Kerala	21347375	25453680	29098518	31841374
14.	Lakshadweep	31810	40249	51707	60650
15.	MP (including Chattisgarh)	41654119	52178844	66181170	81181826
16.	Maharashtra	50412235	62784171	78937187	96878627
17.	Manipur	1072753	1420953	1837149	2166788
18.	Meghalaya	1011699	1335819	1774778	2318822
19.	Nagaland	516449	774930	1209546	1990036
20.	Orissa	21944615	26370271	31659736	36804660
21.	Pondicherry	471707	604471	807785	974345
22.	Punjab	13551060	16788915	20281969	24358999
23.	Rajasthan	25765806	34261862	44005990	56507188
24.	Sikkim	209843	316385	406457	540851
25.	Tamil Nadu	41199168	48408077	55858946	62405679
26.	Tripura	1556342	2053058	2757205	3199203
27.	UP (including Uttaranchal)	88341144	110862013	139112287	174687270
28.	West Bengal	44312011	54580647	68077965	80176197
	India	528585478	658806703	815479910	990922527

Source:

1. Census of India 1971, Series 1, India Part – II – A (i) General Population.
2. Census of India 1981, Series I, India, Part – II – B (i), Primary Census Abstract - General Population.
3. Census of India 1991, Volume 1, Series I, India, Part – II – A (ii) Union Primary Census Abstract.
4. Census of India 1991, Series I, India Part – II – B (i) Volume I, Primary Census Abstract – General Population.
5. Census of India 1991, Series I, India, Part – II B (i), Volume II, Primary Census Abstract General Population.
6. Census of India 2001, Primary Census Abstract (on CD).

Table 8

Here A = 1981

S. No.	Census Years x_i	Population in y	$x = \frac{x_i - A}{10}$	x^2	xy
1.	1951	36.11	–3	9	–108.33
2.	1961	43.92	–2	4	–87.84
3.	1971	54.82	–1	1	–54.82
4.	1981	68.33	0	0	0
5.	1991	84.64	1	1	84.64
6.	2001	102.87	2	4	205.74
7.	2011	121.02	3	9	363.06
		$\Sigma y_i = 51171$	$\Sigma x_i = 0$	$\Sigma x_i^2 = 28$	$\Sigma x_i y_i = 402.45$

From (2), 28

The estimated population in 2021 is given as follows

$$x = \frac{x_i - A}{10} : \frac{2021 - 1981}{10} : \frac{40}{10} = 4$$

Substituting the values of x in $y = a+$

$$Y = 73.1014 + 14.3732 \times 4$$
$$= 73.1014 + 57.4928$$
$$Y = 130.5942$$

The estimated population in 2021 in 130.5942 crores.

First problem: *Source*: Director of Census Operation, Hyderabad.

8. Future Trends in Indian Population : An Assessment

(1) It is seen from Table 4, that the average annual exponential growth rate (per cent) is presented on an increasing scale from 1.04 to 1.64 (2011) but the fluctuations/oscillations are ... progressive and suddenly from 1991 onwards it has been on a decreasing scale recorded from 2.16, 1.97 and further declined by 1.64 perhaps it has been attributed that population has made an indelible mark on average annual exponential growth rate (per cent).

(2) Whereas the progressive growth rate over 1201 (per cent) has started from 5.75 per cent to 407.64 per cent (2011) it is observed that an increase of 80 times per cent to that of 1911 (57.50 × 80 = approximately) if the same trend is continued for 2021, the statistical method of curve fitting holds good besides arithmetic calculation.

9. Conclusion

1. It is concluded that Indian population has been increasing on multiple dimensions such as total population of male population, Female population, literate population and illiterate population, potential population besides senior citizens will be added to the total population of India.
2. It is observed that Government of India, New Delhi should have to take strenuous steps to curb the future population growth on scientific grounds through internet (e-lessons) on over explosion of population and its evil effects on other sectors of development, which should be concentrated from grass – roots level then only, the changing scenario could be implemented from individual to society.
3. Overexplosion of population have its evil effect on human living, food and nutrition, shelter, cloth (MNP), infrastructural facilities and so on. Population education should be designed in the curricula from Primary education to university education, for universalization of population education and proper training and orientation towards desired population and maintain surplus food items to the growing population.

REFERENCES

1. Archana Bhargava, *Resources and Planning for Economic Development, A Case Study* of Sagar Division, MP (pp. 109-116), Pub. by Northern Book Centre, New Delhi, 1991
2. Krishan Chand *et al.*, "*Geographical Analysis of Population Pressure on Land in Maharashtra*", published in *Indian Journal of Regional Science*, Vol. 37, No. 1, 2005 (pp. 138-144), Kolkata.
3. K.L.N. Murthy, Planning for Integrated Area Development - A Case Study from Andhra Pradesh, published by BRPC, New Delhi, 1986, pp. 90-96.

4. Narasimha Murthy, K.L. and M.V.S.N. Murthy, "Population Growth: A Geographical Interpretation, published in *Readings in Development Studies (Socio Geographic Aspects)* published by Print-Well Publishers, Jaipur (India), 1995, pp. 87-114.
5. S.B. Singh, Impact of Urbanization upon the Environment of Varanasi City, published in *Annals*, Vol. 22, No. 2, December 2002, (pp. 74 – 77).
6. Chandna, R.C., *Geography of Population – Concepts, Determinants and Patterns*, published by Kalyani Publishers, New Delhi 2011, p. 61, for details, this may be read by students and researchers.
7. C. Sujathamma, M.V. Reddy and N.B.K. Reddy, City – Rural and Relationship of class I and II towns of Rayalaseema in Vol. 11, Silver Jubilee issue (ed.) By C.R. Pathak, Regional Science Association, India, CK – 134, Salt Lake City, Sector 11, Kolkatta – 700091, p. 119.
8. Surendra Singh and V.S. Chauhan, Regionalization for Rural Development and Planning, published by Independent Publishers Pvt. Ltd., Meerut – 250001 (India), p. 59.
9. Pawan Kumar Sharma, Rural Population Growth Dynamics in Chandigarh Periphery Zone in *Annals*, Vol. 27, No. 1, June 2007, pp. 61-77.
10. *Annals*, Vol. XXX, No. 1, June 2010.
11. *Annals*, Vol. XXXI, No.1, June 2011.

9

Guidelines for Project Report Writing

Project report writing is considered as the third stage of a scientific research project. Generally, the researcher may have to consider seriously about the research topic and the ways and means that have to be identified and recognized for the smooth conduct and completion of his chosen project for his research. The research scholar with his capability, capacity and calibre, should prepare the necessary tools and techniques of his own expression and has to adhere to the application of technology with a coverage of uncovered areas in a time. This chapter ninth deals with the a model project report writing and 9.2 is research proposal. The research project may be divided into five stages. They are:

(1) Introduction, (2) Survey of Literature, (3) Data Collection, Processing and Analysis, (4) Major Findings of the Study and (5) Bibliography

According R.P. Misra, "Scientific Writing" has been defined as that writing which is directed to scientists outside the authors speciality and to the general public. In both scientific and technical writing the writer should strive always from accurate, logical, comprehensive and smoothly readable expression of facts and ideas".

An attempt is made by the author in this chapter to present the stages of a research project writing. They are:

(1) Organizing the disorganized writing;
(2) First draft of the Manuscript;
(3) Writing for the awaiting young researchers;
(4) Style and functional English;

(5) Second draft and revision;
(6) Completion of the manuscript; and
(7) Format for scientific report writing.

Students of research need proper training for the preparation of their course work/assignment particularly at the post graduate level and M.Phil level, deals with research papers, dissertations and a thesis. The important aspects in this respect should have to be followed by the P.G. Students level including Ph.D. Scholar invariably in different angles. They are:

(1) Organizing the disorganized writing: In general, the students should learn to earn the necessary skills for preparing the term papers as well as partial fulfilment for the award of degree by a University, i.e., M.A./M.Sc./M.Tech. by the Indian Universities. A prescribed course of research activity by way of formulating the ideas into a standard format and bring them into concrete form in order to identify and find a solution to the research problem.

A student is expected to equip and accommodate himself of the concept/core of the subject when he is with his teacher at the post graduate level, i.e., hearing the lecture and writing down the important aspects during the length of the lecture. Later on, the students have to prepare the necessary notes from reference books and text books. This is a way of preparing comprehensive notes for examinees by themselves.

The material gathered or collected in the library/standard reports by the student is in a form of disorganized state. This material should have to be put in a cogent manner in order to have a comprehensive look and preparing thesis in a complete manner for assessment or for adjudication. Sometimes, the student may write the material in different ways in order to comprehend the subject. The complexity should have to be surpassed by way of putting them in an organized way with utmost care without leaving any aspect of it.

(2) First draft of the manuscript. The preliminary way of analysed data with statistical treatment along with basic maps which help geographic cartographic methods for power point presentation or publication/seminar/conference as a conceptual presentation. The basic idea is presented in a skeletal form and the firm ideas are not at all presented in this report. Data processing by different methods are expressed and later on exposed before the learned council, i.e., in a seminar. Adding the advices, important points at different stages by

different people (which are offered) should have to be obediently, faithfully, and simultaneously should adhered to add to the skeletal form. The above format gives an idea and helps to add the necessary and important ingredients to get a good fame and design of reference to the basic report.

(3) Writing for the awaiting young researcher: The concepts may be presented in different forms by the author. The authenticated report form is much a wanted format. Generally, students are expected to become experts in their own traits. To become experts in the subject, the latest arrivals in the subject and subject related topics are to be represented and presented intensively when the student is incapable of getting material and capable of mixing up to the existing material, still he awaits for further material and guidance from the supervisor. Keeping in view of their site/situation, a seasonal writer/researcher's research may indeed help him/her in the furtherance of research. A consultant and a seasoned researcher will prepare the latest material with techniques of science and technology in the subject and the researcher's have to gear up to the ideas besides of their own, in implementing the ideas as an addition which really helps as a flesh and blood to the skeletal format of the first draft of the manuscript, which is like heart functioning.

(4) Style and functional English: The researcher should be able to identify and recognize the needs of the seciety with the help of science and technology, man may be able to fulfil the aspirations of the society. The essentials or a, b, c of style are defined as *accuracy,_brevity* and *clarity* while presenting the concepts in a concise form. The style of a composition is the selection of words, structural pattern of compound and complex form besides simple structures, the ideas are to be expressed, pin pointed patterns of attitudes at a high altitude and the audience addressed will help the entire text of the manuscript which enriches and highlights the subject more complementary rather than competitive ideas to the reader while grasping the subject by the readers.

(5) Second draft and revision: The earlier draft is a skeletal report in which contains the study, wherein the researcher may be able to place it in its general scientific context to adjudge the adequacy/or otherwise (improvement at a later date) of its methods and it should be able to form an opinion of how seriously the findings are to be taken or to what extent these should be dependent and act as guide to the future scientific action to be taken up by scientific communities so for the development of science and art of geography.

The research report must satisfy the following points in order to supply the necessary information to the researcher social scientist:

1. Statement of the problem with which the study is concerned.
2. The research procedures:
   (i) The study design,
   (ii) Independent variable and dependent variable,
   (iii) The universe and the nature of the sample,
   (iv) The data collection techniques, and
   (v) Method of statistical analysis,
3. The results of the study/major findings
4. The implications of the research results and thereby practice.

1. Statement of the research problem for doing/conducting research:

(1) The first step in making a statement of the problem and concerning area. For example, Impact of rainfall on agricultural productivity in a drought prone area, Andhra Pradesh.

2. The Research Procedures:

1. *The Study Design*: The above study may be conducted at grass roots level/mandal level/district level, i.e., Micro level (village level); Macro level (group of villages forming a mandal/groups of mandals forming a *macro region*, i.e., the study is related to micro level, meso level, or macro level.
2. *Independent Variable*: Rainfall (seasonal-break-up of rainfall, i.e., daily rainfall, weekly rainfall, monthly rainfall) include dependent variable. Crop production and productivity levels.
3. *The Variable and Nature of Sample*: Deciding the pattern of rainfall/drought proneness, the study should be decided to a level, which satisfies the criteria and may be able to explain and express the nature of the pattern (production trend by the method of moving average helps to map the seasonal trend) or seasonal indices.

   Statistically, the sample design should have to be considered for drawing the sample to conduct the above study in a specified time frame keeping in view of the feasibility of the individual as well as financial resources, size of the sample ranges from 5 per cent to 10 per cent representing the universe.

4. *Data Collection Techniques*: Statistical methods of sampling should have to be adopted like simple sampling/random sampling/systematic sampling in the study representing the universe.

   Standard questionnaire have to be designed in order to conduct the survey in the study area. Primary data and secondary data have to be collected relating to the study area and the drought effect should have to be identified upon people/animal and agricultural holdings and so on (operations).

5. *Method of Statistical Analysis*: Statistical method should be explained prior to the analysis. The second step is to carry out the statistical exercise by putting the collected data for processing. (Collected data is considered as input) and analyzed data (result is considered as the research output).

(c) The Results or Major Findings

The generated statistical output should have to be conducted positively/negatively to accept the result. The rainfall incidence upon agricultural productivity is positively correlated at 5 per cent level or 1 per cent significance level and the application of null hypothesis.

The scientific reader is also concerned with the analysis of data. He needs to be informed about the techniques used to determine the impact and its analysis which differs from normal to particular characteristic phenomenon. It is used to simply name the technique used (whether it is correlation coefficient method/principal component method) and to indicate the level of confidence the researcher had accepted in deciding whether characteristic feature and its incidence are to be considered significant.

The basics in presenting the findings are to give all the evidence relevant to the research-question, whether or not the results are in accordance with the researcher's views. This rule of scientific reporting is a must. "The scientist is not free to use of his own tastes, unless and otherwise, it is statistically sound and acceptable to statistical tests. The decision as to what is relevant is guided mainly by the research-problem and hypothesis (if any) with which the study is concerned". (Quotation mine please).

In an exploratory study, for example, the content and organization of the report would be much less clearly suggested by the study design and analysis of data. Generally, the researcher must depend more on his own judgment in deciding the outline of his report. Further, it is warranted on the part of the researcher that he should state it clearly the problem with which he is concerned, the procedure by which he worked upon the problem, the conclusions he arrived at and the basis for his valid conclusions". (Quotation mine please)[1]

The following references are useful guides to assist the writer when he is in doubt. Therefore, it is suggested that the following books or one of a similar nature should be available to the researcher/ writer.

1. Webster's New Collegiate Dictionary.
2. Oxford Dictionary (for its many examples of correct use of words and phrases).
3. Soule's Dictionary of English Synonyms.
4. Encyclopaedia of Geomorphology.
5. Dictionary of Geography (Physical Geography) by Moore.
6. Dictionary of Geography by F.J. Mank House.

Completion of the Manuscript

The following are the steps, which are proper to follow them precisely for concise completion of the manuscript:

1. Order of development of the argument and see whether the writing is understandable when it is read rapidly aloud.
2. Particular attention on the aspects of the material such as (1) the introduction, (2) the observations, and (3) the conclusions.
3. Experimental data seldom occupy the central or much part of the manuscript. Too condensed explanation leads to deterioration of originates to avoid this pattern, we have to develop explanation in its true form and enriched originality offers more validity and validating points/score to accept it (text and argument). While analyzing the data and writing the assignment or thesis, we should have to consult a senior colleague both in the subject as well as a language professor

to meet the requirements of a good and balanced report. If it is possible to the researcher, he should try to read the thesis/ assignment before him to adjudge the merits and demerits of its validity.

4. Summary consists of few paragraphs, emphasizing the importance of the problem/subject and the main conclusions arrived/reached by the writer.

Finally, it is concluded that "the complexity of the material as well as the care in the writing once determined when one can stop revising and pronounce the most recent draft the last".[2]

Bibliography

Bibliography is the last important part of a research report. A bibliography is nothing but the sources used in the preparation of the report. These may be 1. Books, 2. Articles in periodicals, 3. Reports and documents issued by both official and unofficial agencies, 4. Interviews, 5. Lectures, 6. Radio and television tasks, 7. Computer Graphics and later on monographs are to be listed.

Final Touch

Good work sometimes needs final touch by the researcher/author. So, it should be edited before it is sent for typing. The illustrative material should be mounted, folded and inserted at appropriate places.

"The preparation of a scientific report is different in one fundamental respect from an argumentative essay or a literary effort: the self conscious presentation of negative and positive evidence, along with the techniques by which that evidence is obtained. The scientific report is expected to become part of the cumulative body of verifiable knowledge, it must be presented so that its data can be repeated, its logic followed and its significance ascertained. Further, it is for the investigator who is unwilling to face both the strength and the weakness of his work will soon have brought to his attention by others. It is thus which puts such harsh demands upon the researcher, but thereby is created the only possible basis for developing, accepting and utilizing the conclusions of this young, exciting science".[3]

REFERENCES

1. Wilkinson and Bhandarkar, *Methodology and Techniques of Social Science Research*, pp. 389-395, Published by Himalayan Publishing House, Bombay-400004, 1982.
2. Misra, R.P., *Research Methodology: A Handbook*, pp. 52-69, Concept Publishing Company, New Delhi-110 059, 1988.
3. Anderson and *et al.*, *Thesis and Assignment Writing*, Wiley Eastern Limited, New Delhi, 1970.

9.1

Research Proposal

Nutritional Deficiency Diseases and their Prevalence : A Case Study of Urban Pondicherry

Smt. G.V. Kameswari and Dr. K.L. Narasimha Murty

Health status is one of the important measures of quality of population. Good health status of the citizens in an economy determines the economic development of the country. India has succeeded over the last six decades in doubling life expectancy, reducing infant mortality and child mortality rates. Yet the percentage of number of people who are below poverty line being more than 30 per cent and are malnourished, illiterate and in bad health. (World Bank 1998)

Despite efforts to combat malnutrition among the population, particularly among those belonging to low socio-economic groups in the rural and tribal areas and urban slums, malnutrition continues to be an important public health problem in India even after 60 years of independence. It is estimated that a majority of the undernourished people in the world live in the Indian subcontinent. An assessment of the actual situation of the diet and nutritional status of the population, and the changes that may have taken place over time is important for planning appropriate strategies to mitigate the suffering of the community. This is particularly important in the context of the nutrition goals set in the National Nutrition Policy of the Government of India.

1. (Indian journal of Medical Research 1998, by Vijayaraghavan. K. Rao, D. Hanumantha Rao).

Lincoln C. Chen (1987) compared the health status and healthcare systems of China and India, and developed a conceptual framework on the determinants of the health status of population taking into account both the performance of health care system and non-medical factors that include health. The paper concludes that the economic crisis presents an opportunity for a structural view of the society's health system and that health policy options cannot be confined to health care sector only but should be included as an element in the socio-economic policy option package. Only then it will be possible to minimize the adverse impact of economic crisis on health, nutrition and social welfare of the disadvantaged groups.

Earlier studies demonstrated that in the communities which depended mainly on cereals and millets, the primary bottleneck, in the dietaries of the pre-school children, is energy deficit or food deficit. These and other studies helped in formulating the scientific basis for the supplementary feeding programmes in operation in the country.

The above macro level analysis explains the necessity for micro level analysis of the problem and implementation of National Policies at the grass root level. An investigation on the socio-economic and demographic determinants of nutritional deficiency diseases growth rate will be useful to identify the necessary policy options to reduce the problem to the minimum. Some of the important determinants that can be taken are the per capita income, literacy rates, urbanization, density of population, percentage of population below poverty line, percentage of people not regularly exposed to media, work participation rate of females and per capita health expenditure.

The two municipalities of Urban Pondicherry are Pondicherry and Ozukarai. The present study is aimed at analyzing the Nutritional Deficiency Diseases in these two municipalities and this urban area of Pondicherry will be taken as a case study. Population growth rate from the year 1901 to 2001 will be calculated and analyzed whether it is linear or exponential and will be projected to the future date. Its impact on the food consumption will be identified. The projected population structure can also be included in the planning and implementation of the nutrition policies. Field studies will be undertaken to know and analyze the different determinants of the nutritional deficiency diseases in the below poverty level groups of people. Being a comparatively small area for administration and its advantage of being a Union Territory, the effectiveness of implementing development programmes laid down by the

Government is expected to be high and the role of Self-Help groups in this context will be studied. Pondicherry Government is currently taking full care of the economically, caste and class-wise disadvantaged sections of society. Government provides financial assistance to these sections whenever natural calamities strike. The recent rehabilitation programmes after tsunami occurrence is a good example. Pondicherry urban area has a highly equipped Government Hospital and internationally renowned central government institution JIPMER with highly motivated and efficient medical personnel and government is providing an excellent medical care. There are nutritional experts in these two institutions. In addition four private medical colleges are situated in Pondicherry. There also nutritional experts are available. Their help will be taken to list and assess the nutritional deficiency diseases and the counselling to be provided for the remedies and their implementation to the economically weaker sections by Government and Self-Help groups will be studied. Government is already providing nutritional food to the school-going children at government schools. Its implementation will be assessed.

The most common nutrition deficiency diseases prevalent in this area and their growth or decline over a period of time will be studied. The effectiveness of the Government policies and steps taken by government in eradicating nutrition deficiency diseases will be determined and analysed. A survey conducted by the Central and State Health Directorates. Indian Council of Medical Research and Medical institutes have clearly demonstrated that not even a single State or Union Territory is free from the problem of Iodine Deficiency Diseases. No data was available from the Union Territory is free from the Pondicherry on the prevalence of iodine deficiency. (Nutrition Foundation of India 1994, 15). The nutritional levels of the food intake of the poor will be estimated as per the standards laid down by the National Institute of Nutrition, Hyderabad which has attained global recognition for its studies on various aspects of nutrition. One of its objectives is to identify various dietary and nutrition problems prevalent among different segments of population and to evolve effective methods of management and prevention of nutritional problems. These will be undertaken at the micro level in this case study and interpreted and suggestions will be put forward for implementation.

REFERENCES

F.H. Harbiston, "Human Resources in Development Planning in Modernizing Economies." H.R. May, 1962.

Norman L., Hicks and Paul P. Streeten. "Indicators of Development. The Search for a Basic Needs, Yardstick" *World Development*, Vol. 7, 1979.

M.L. Jhingan, "*The Economics of Development and Planning*" Vrinda Publications

World Health Report (1998), *Health Across the Life Span*, pp. 60-64.

Lincoln, C.C. (1987). Cropping with economic crisis: Policy Development in China and India. *Indian Journal of Medical Research 1998*, by Vijayaraghavan K. Rao, D. Hanumantharao.

Vir S. Control of Iodine Deficiency. The National Programme Current Status. *Bull Nutrition Foundation of India*, 1994, 15.

Bibliography

A.N. Strahler, *Physical Geography*, Wiley Eastern, New Delhi.

Bhat, L.S. (1978), *Micro Level Planning: A Case Study of Karnal Area*, Haryana, India.

Chandna, C. (2011), *Geography Population*, Kalyani Publications, New Delhi.

Crow, P.R. (1933), *The Analysis of Rainfall Probability*, Scottish Gorg, Magazine, Vol. 49, pp. 73-91.

Freeman, T.W. (1968), *Geography and Planning*, London, Hutchinson University.

GOI, Planning Commission (1965), *Resource Development Regions and Divisions of India*, New Delhi.

Haggett, P. (1959), *Locational Analysis in Human Geography*, London, Arneld Press.

Hauser, P.M. (ed.) (1963), *The Population Dilemma*, Englewood Cliffs, Prentice Hall, Inc.

Higbel, E. (1958), *American Agriculture: Geography and Resource Conservation*, New York, John Wiley & Sons.

Hussain, M. (1972-1973), *Crop Combination Regions of Uttar Pradesh: A Study in Methodology*, GRI, Vol. 34-35, pp. 134-156.

Joseph Hutchinson, (Ed.) (1969), *Population Food Supply, Essays on Human Needs and Agricultural Prospects*, Joshi.

Learmonth, A.T.A. and L.S. Bhat, (eds.) (1960), *Mysore State: Atlas of Resources*, Vol. 1, Indian Statistical Institute, Calcutta.

Murty, K.L.N., *Indian Social Science Research* (2012), Teja Publishers, Hyderabad (pp. 47-69).

Murty, K.L.N., *Applied Geo-environmental Studies* (2008), Vol. 1, Mangal Deep Publications, Jaipur (pp. 25-54).

Murty, K.L.N., *Applied Geo-environmental Studies* (2008), Vol. 2, Mangal Deep Publications, Jaipur (pp. 67-81).

Murty, K.L.N., *Applied Geo-informatics*, (2007), Mangal Deep Publications, Jaipur, (pp. 152-161).

Murty, K.L.N., *Ecology, Environment and Development* (2003), Mangal Deep Publications, Jaipur (pp. 173-180).

Murty, K.L.N., *World Geography*, (2003), Telugu Academy, Andhra Pradesh, Hyderabad (pp. 67-152).

Murty, K.L.N., *Geographical Research*, (1999), Concept Publishing Company, New Delhi (pp. 110-134).

Murty, K.L.N., *Readings in Development Studies (Socio-Geographic Aspects)*, Print Well, Jaipur, 1995, (pp. 115-137).

Murty, K.L.N., *Research in Geography*, 1992, Ashish Publishing House, New Delhi (pp. 37-72).

Murty, K.L.N., *Planning for Integrated Area Development* (1986), BRPC, New Delhi (pp. 102-133).

Murty, K.L.N., *Regional Geography of India* (2012), Telugu Academy, Hyderabad, (in press) (pp. Urbanization in India).

Murty, K.L.N., *Manachitra Mulyankanams*, Telugu Academy, Andhra Pradesh, Hyderabad (2010), (pp. 37-47).

Narasimha Murty, K.L., "Spatial Variations in Agriculture" (pp. 117-120), *New Dimensions in Geography and Allied Sciences*, Ed. by R.S. Dixit *et al.*, The Institute of Geographers, India, Lucknow-226006 (India.).

Narasimha Murty, K.L., and P.V. Sarma, Central Place Theory in Indian Contest—A Survey, *The Deccan Geographer*, Vol. 25, Jan.-June, 1987, No. 1 (pp. 151-158).

Narasimha Murty, K.L., and P.V. Sarma, Identification of Central Places in Tanuku Taluk, *Indian Journal of Regional Science*, Vol. 27, No. 2, 1985 (pp. 20). Published by Regional Science Association, India, Indian Institute of Technology, Kharagpur, India.

Narasimha Murty, K.L., and P.V. Sarma, Animal Husbandry Resources and Poultry Farming Resources in Tanuku Taluk, A.P. India, Published in ACARSH, Belgaum (1985), p. 11.

Narasimha Murty, K.L., and P.V. Sarma, Environment and Agriculture: A case study of Krishna Delta Region in Andhra Pradesh. Ed. by H. Ramachandran, "Environmental Issues in Agricultural Development" (1990), Concept Publishing Company, New Delhi (pp. 95-104).

Narasimha Murty, K.L., *Land-use Planning Model for Srikakulam and Environs: An Urban Dynamic Process in a Rural Setting* (1992), 27th International Geographical Union, Washington, D.C., U.S.A.

Nath, V. (1966), *The Growth of Indian Agriculture*, G.R.I. 59 (pp. 348-372)

O. Riordan, T. (1971), Perspectives of Resource Management, London, Pion Ltd.

Royen, W, Van (1954), Agricultural Resources of the World, Prentice Hall.

Sengupta, P (1966), Planning Regions for Resource Development in India, *National Geographic Journal*, India, Vol. 12, No. 1 (pp. 1-23).

Singh, R.L. (1971), *India: A Regional Geography*, R.L. Singh, Ed. BHU Varanasi.

Singh, R.Y. (2010), Geography of Settlements, Rawat Publications, Jaipur.

Taylor, G. (Ed.) (1951), *Geography in the 20th Century*, New York.

United Nations, Department of Economic and Social Affairs (1964), *Population Growth and Manpower in Sudan*, New York.

Economic and Social Affairs (1964)

Wise, M.J. (1969), *On the Utilization of Resources, Geography*, Vol. 54, pp. 257-70

Young, A. (1978), *Recent Advances in the Survey and Evaluation of Land Resources, Progress in Physical Geography*, Vol. 2, pp. 462-79

Zinmerman, E. (1951), World Resources and Industries, New York, Harper and Row.

Zimmerman, Leonard (1962), *An Economic Historical View of National Resource* – Use and Conservation.

Index

Accuracy of results, 57
Adoptive capacity project, 91
Aerial photo elements, 33
Aerial photographic method, 89
Agra University, 151
Agricultural and Industrial Geography, 81
Agricultural development, 4, 6, 8, 85
Agricultural Efficiency, Measurement of, 104
Agricultural modernization, 8, 11
Agricultural prosperity, 7
Agricultural resources, identification of, 8
Agricultural situation, 6
Agricultural technology, 96
Agri-labourers, 120
Agro-sector, 6
Agro-settlement system, 5
Air pollution, 89–90, 157, 159
Air quality, 5
Alluvial soils, 19
American Geographical Society, 89
Analysis and Interpretation of Data, 30
Animal husbandry, 113
Area of Principal Crops Irrigated (APCI), 103
Area under Principal Crops (APC), 103
Arithmetic Method, 195
Association of American Geographers, 15, 79
Atlas of Diseases, 89
Atlases of Scheduled Castes, 112
Autonomous organizations, 36
AVC GIS software, 161

Behavioural patterns, 12
Beiger Institute of Ecological Economics, 79
Biodiversity, 3, 7, 61, 85
Biological processes, 2
Biotic factors, 95
Bureau of Business Research, 15

Case study method, 69, 74–75
merits of, 74
Cash crops, 101
Caste factors, 39
Causative factors, 25, 61, 64, 89
Census method, 53, 55, 57
Census of India, 29, 32, 115, 117, 138, 141, 146
Census Operation, 198, 201
Central Ground Water Commission, 91
Central Place theory, 15, 98, 126–127, 131–132
Centrality index, 128, 132, 136
Centrality, measurement of, 128
Centre for Regional Studies, 155

Centrographic analysis, 109
Child labour, 12
Choropleth method, 142
Climate change, 4–5, 91, 163–164
Climatic factors, 17
Climatic impact, 17
Climatology, 1, 4, 17, 19–20, 82, 84–85, 152, 157–158
Coastal vulnerability index, 93
Coefficient
 of correlation, 138
 of development, 142–143
 of variation, 90
Commission for Asia and Pacific, 79
Commissioner for Public Institution, 146
Complementary methods, 25
Component method of factor analysis, 103, 150
Composite development index, 146
Composite index method, 146
Composite index, 100, 145–147, 149
Compound Annual Growth Rate (CAGR), 192
Computer Assisted Cartography (CAC), 156, 161
Concentration index techniques, 191
Conflict-political economy, 7
Confluence zone, 85
Controlled Observation method, 68
Cooperative credit society, 144
Corporate index, 107
Correlating factors, 25
Correlation, 21, 92, 101–102, 137–138, 145, 178–180, 193
 coefficient, calculation of, 175
 matrix, 148–149
Cost factor, 132
Crop associations, 107
Crop concentrations, identification of, 108
Crop diversification, 108–109
Crop intensity, 97
Crop patterns, 4, 93, 95
Crop planning, agricultural, 4
Crop production, 8, 207
Crop productivity, 8, 95
 agricultural, 4
Crop ranking, 101–102
Crop-combination, 101–102, 108–109
Cropping Intensity (CI), 95–96, 136
Cropping intensity index (CII), 107
Cropping patterns, 7, 94–95, 101–102, 107–108
Cultivated land, 121, 189
Cultural factors, 14, 95, 130
Cultural variables, 32–33
Curve fitting, 122–123, 188, 190, 193, 197, 202
Curve fitting formula, 196

Davies method, 128
Decennial growth of population, 116
Decentralization policy, 12
Declarative forms, 45
Declarative hypothesis, 45
Demographic dynamism, 112
Demographic index, 103, 149
Dense forest, 140
Department of Economics and Commerce, 78
Desert National Park, 3
Detailed Projects Report (DPR), 92
Development
 coefficient of, 144
 composite, 142
 cultural, 14
 global, 7
 haphazard, 190
 infrastructural, 126, 189
 instantaneous, 11
 integrated, 148, 161
 irrigational, 107
 low level of, 142
 optimum, 8
 pasture, 7
 planned, 63
 regional/national, 9

semi-urban fringe, 10
spatial, 10
strategic, 8, 11
technological, 107
Developmental planning, 126
Differential index, 137
Digital image processing software, 163
Direct contact method, 80
Disadvantaged groups, 213
Disasters, 3, 162
Disparities, 8–10, 13–14, 102, 145
of development, 14, 149, 157
regional, 6, 8, 12, 81, 102, 104, 145
Documentary method, 69
Doi's method, 108
Drainage, 88, 162–163
Drainage basin, assessment of, 4
Drainage maps, 86, 88
Droughts, 4, 11, 162–163
Drought prone areas, 5, 207
Dry agriculture, 7

Ecological model, 164
Ecology of Malnutrition, 89
Economic
development, 8, 78, 190, 212
environment, 136
factors, 14, 39, 95
geography, 5–6, 81, 83, 94–95, 151, 157
reforms, 8
resources, 6, 148
Education weightage, 133
Empirical tests, 48
Employment generation, instrument of, 6, 12
Environment
influence of, 101–102
quality of, 148
Environmental degradation, 101
Environmental factors, 39
Environmental Impact Assessment (EIA), 92, 164
Environmental Management Plan (EMP), 93
Erosion, 2
Ethnic population, 7, 12
Exclusive factors, 49
Experimental groups, 64
Experimental method, 24, 43, 52–53, 61, 64–65, 68
demerits of, 64
Exploratory design, 49
Export markets, 99
External factors, 62
Extraneous factor, 62

Factor analysis, 101–102, 111, 148–149
Fertility, 111
Field study, 24, 87
Five Year Plans, 6, 13
Floods, 4–5, 11, 85, 162–164
Flood Hazard Mapping (FHM), 164
Flood prone area, 5
Food availability, 99–101
Food insecurity, 99
Food Security Standard Statistical Method, 101
Food security, 7–8, 99–101
regional Imbalances in, 99
Food stability, 99–100
Foodgrains availability, 121, 189
Forest plantation, 140
Forest Resources Management, 163
Forest Survey of India (FSI), 163
Fundamental Research, 22

Generalizations, 27, 31, 40, 42, 46, 55
Genetic factors, 44
Genetic method, 25
Geochemistry and Geomorphological methods, 156
Geographic association, 10
Geographic cartographic methods, 205
Geographic distribution of animals, 20
Geographic Information System (GIS), 3, 16, 93, 98, 102, 130, 155–156, 160–163

Geographic information, 22–23
Geographic Positioning System (GPS), 16, 155–156
Geographic research methods, 23, 68
Geographic research, design of, 48, 68
Geographic survey, 48, 68
Geographical geomorphology, 20, 154
Geography laboratory, 53, 65–66, 68
Geography
 branches of, 11, 14–16, 68, 81–83, 95, 127, 131, 154–155
 development of, 18, 110, 153
Geo-informatics institute, 160
Geologic phenomena, 20
Geomorphology, 1–2, 17, 19–20, 24, 31, 82, 84–85, 152, 154–155, 157, 209
Geo-social phenomena, 38
Gibbs-Martin Method, 108–109
GIS software ILWIS, 102
GIS techniques, 86–88, 102, 163
Godlund's Method, 128
Good questionnaire, characteristics of, 73
Gravity Model, 98
Grazing land, 96
Greater Hyderabad Metropolitan Corporation area, 121
Grid system of random sampling, 60
Groundwater samples, collection of, 158
Group interaction, 80
Group research, 52
Growth centres, 16, 153

Health Care Delivery Centre (HCDC), 160
Health Care Facilities (HCF), 13, 159
Health care system, 13, 158–160, 213
Heeday's land-use model, 98
Heterogeneous factors, 62
Hierarchical methods, 25
Hierarchical orders, 132, 140
High Yielding Variety (HYVs), 97
Hill slope profiles, 3
Homogeneity, 54
Homogeneous groups, 55
Household industries, 113, 120
Human poverty index, 9
Human resource development, 8–9, 141–142
Hypothesis formulation, 43

Index
 of Agricultural Efficiency, 105
 of Agricultural Production, 117
 of Diversification, 108
 of pressures, 122
India Soil and Land Use, 32
Indian Council of Medical Research and Medical Institutes, 214
Indian Institute of Public Administration, 78
Indian Institute of Tropical Meteorology (IITM), 90
Indian Meteorological Department, 32
Indian River System, 85
Indian Space Research Organization, 153
Indirect contact method, 80
Industrial activity, 151
Industrial development, 6, 8
Industrial Geography, 95, 151, 157
Industrial landscape, 151
Industrialization, 6, 121
 high, 142
 Mapping, 78
Industries weightage, 134
Inequalities, 10, 12, 193
 development, 157
 regional, 8, 10
 social heterogeneity/economic, 54
Infra index, 103
Innovative society, 11
Institute for Social and Economic Change, 78
Intensive efforts, 27–28, 44–45, 56
Inter-census population, calculation of, 188

International Geographical Union, 98
Interview method, 26, 69, 79–80
Irrigation, 85, 87, 91–92, 97, 104

Karl Pearson's Method, 178
Kendal's rank order score method, 104

Laboratory method, 25
Labour markets, 7, 12
Land capability, 86
Land degradation, 3
Land management programmes, 85
Land reorganization, 129
Land slides hazard, 3
Land transformation, 8, 129
Land utilization
  pattern of, 118
  statistics, 97
Landform structure, 20, 31
Land-man ratio, 12
Land-use
  categories, 96–97
  changes, 96
  dynamism, 139
  intensification, 8
  pattern, 4, 96, 121, 131, 139
Law of Retail Gravitation, 15
Law of statistical regularity, 55
Library method, 24–25, 69, 75–76, 77
Literature survey, 27, 39, 75, 204
Livelihood system, 12
Lorenz curve and Ginni's method of coefficient, 192
Lottery Method, 60
Lower plain land, 85

Madras Institute of Development Studies, 78
Malnutrition, 89, 212
Mapping, 3, 18, 25, 30–31, 65, 87–89, 129–130, 145, 154–155, 163
Marginal workers, 115–116, 120
Market forces, 96
Medical care system, 13
Memory, 138
Methodological approach, 15
Methodological field, 111
Micro level planning, 8, 153
Migration, 11–12, 62, 111, 157
Minimum cost factor, 8, 33, 95
Minimum deviation method, 108
Modified factors, 86
Morphometric analysis, 86, 88
Morphometric units, identification of, 86
Mountain development, 82
Multi stage sampling, 53, 60–61
Muslim Rule, 151

National Association of Geography, 79
National Atlas Organization, 112
National Federation of Co-operative Sugar Factories Ltd., 109
National Geophysical Research Institute, 32
National Institute of Nutrition, 32, 214
National Institute of Rural Development, 77
National Institute of Urban Affairs, 78
National Nutrition Policy, 212
National Population for National Count, 32
National Remote Sensing Agency/Organization, 153
National Remote Sensing Centre (NRSC), 32–33
National Sample Survey (NSS), 32
Neo-systematic format, 36
Net sown area, 90, 96–97, 104–105, 143–144
New ecological system, 22
Non-medical factors, 213
Non-participant observation, 70–71
NSS Report, 115
Null forms, 45–46
Null hypothesis, 45, 208

Observation Method, 68–69

Occupational structure, 6–7, 12, 127, 151
Oceanography, 1, 17, 19, 82, 84, 152, 157
Operational framework, 162
Operational techniques, 65

Paddy, 86, 93, 97, 186
Participant observation, 70–71
Periodic market system, 15
Periodical surveying, 63
Photogrammetric methods, 156
Physical Quality Life Index, 148
Physical variables, 33, 42–43
Plane table survey, 65
Pollution Control Board, 160
Population density, 91, 144, 190–191, 213
Population education, 202
Population pressure, 12, 96, 121, 189–190
Powerful strategies, 7–8
Pre-independence era, 191
Primary data, 24–25, 30–33, 36, 68, 70, 75, 77, 90, 208
Primary education, 202
Primary Health Centre, 142
Production of Ungauged Basin (PUB), 3
Public hearing, 93

Qualitative method, 23, 150
Qualitative tests, 23
Quantitative method, 24, 141, 150, 152
Quartile methods, 106
Questionnaire, 33–35, 49, 51–52, 72–73

Random sampling, 52–53, 60
Rank Differences method, 138
Rationalization, 126
Ravine land, 3
Regional development, 6, 8–9, 12, 14, 81, 141, 145, 148, 157
Regional domestic crops, 101–102
Regional Science Association, 15
Remote Sensing Methodology/Aerial Photo Interpretation, 153
Remote Sensing Organization, 153
Research
  design, 27, 29, 48–51
  hypothesis role in, 41
  informatics, 35
  methodology, 81, 90
  problem, 27–28, 30–31, 36–37, 39–40, 49, 69–70, 205, 207
Resource endowments, 6
River basins, 85, 88, 140, 152, 164
Rock structure, 19, 24
Rural development, 10, 15, 77, 79, 81
Rural Electric Service Connection for Agriculture (RESCA), 103
Rural Non-Farm (RNF), 12
Rural population, 9, 13, 138, 142, 192
Rural poverty, 15
Rural settlements, 11, 127, 130–132, 191
  hierarchy of, 132
Rural spatial settlement system, 15

Sampling, 53–55, 57, 60, 208
  Definition of, 54
Scale of Index value, 143
Scarcity index, 159
Schedule method, 35, 68, 72–73
Scientific explanations, 18
Scientific method, 18–21, 41, 57, 64, 150
Scientific organizations, 153
SD formula, 106
Self-Help Groups (SHGs), 214
Service centres, 16, 130–132, 153
  Hierarchy of, 131
Settlement system, 5, 12–13, 15–16, 33, 125–127
Sex ratio, 120, 184
Slope analysis, 63, 140
Slum population, 117

Social development, 78
Social research, 21–22, 38, 69
Social science research, 20, 23, 26, 72, 77
Social sector, 8–9, 11
Socio-economic
factors, 101–102
survey, 64, 164
technology, 54
variables, 33, 137
Soil
deltas/red soils/black, 19
erosion, 94, 163–164
pedalogical factors, 7
pollution, 157
productive, 86
quality, 20
samples, 24, 66
Sopher's Disparity Index, 137
Space technology, 162
Spatial development strategy, 9
Spatial distribution, 20, 110–111, 139, 143
Spatial medical care system, 13
Spatial organization, 131
Spatial patterns, 15, 137, 159, 164, 189
Spatial variations, 9, 38, 95–96, 101–102, 130, 157
Standard deviation (SD), 100, 106, 110, 144, 147–148, 178
Standard deviation grouping techniques, 146
Standard deviation method, 147
State Pollution Control Board, 90
Static temporal model, 126
Statistical methods, 24, 43, 54, 188–189, 193, 202, 207–208
Statistical tests, 208
Statistical treatment, 205
Structural development, 157
Students T-test technique, 101
Suitable alternative index, 108
Survey, 28, 33–35, 57, 64, 90, 147, 208, 214
Survey method, 25
Survey of India maps, 140
Survey of India sheets, 86, 88
Survey of India toposheet, 91
Surveying instruments, 63, 65
Surveying methodology, 156
Swamp lands, 85
Systematization, 27, 30

Tabulation, 27, 30, 73
Technological inducement, 96–97
Temporal land-use, 97
Thematic mapping organizations, 14
Theory formulation, 111
Thermodynamic system, 62, 98
Thuhuen's land-use analysis, 131
Time-space elements, 38
Tippets Numbers Method, 60
Total Cropped Area (TCA), 96, 103, 108
Transportation, 2, 157

Underground water, 2, 19, 157–158
Unemployment survey, 113
Urban dynamic model, 98
Urban planners, 9–10
Urban settlements, 9, 33, 125, 127–128, 130, 140
hierarchy of, 140
Urban space, 9, 128
Urban systems, evolution of, 127
Urban zone, 9
Urbanization, 6–7, 9–10, 100–112, 121, 129, 138, 190, 192, 213
impact of, 190
level of, 138
Urban-Land differentials, 136

Vegetation, 82–83, 92, 163
Verbal survey, 34
Veterinary Centres, 103
Village anatomy model, 126
Vulnerability index, 91